AF531687

POTENTIALS OF NANOTECHNOLOGY

ENCYCLOPAEDIA OF NANOSCIENCE

Vol. IV

POTENTIALS OF NANOTECHNOLOGY

By
Dr. S.K. Prasad
School of Studies of Zoology & Biotechnology
Vikram University
Ujjain

DISCOVERY PUBLISHING HOUSE
NEW DELHI-110002

First Published-2007

ISBN 978-81-8356-247-8 (Set)

Published by

DISCOVERY PUBLISHING HOUSE
4831/24, Ansari Road, Prahlad Street,
Darya Ganj, New Delhi-110002 (India)
Phone: 23279245 • Fax: 91-11-23253475
E-mail: dphbooks@rediffmail.com
dphtemp@indiatimes.com

Printed at:

Sachin Printers, Delhi

PREFACE

There has been rapid progress in nanoscience over the last few years, particularly in the area of miniaturization. The present title **Encyclopaedia of Nanoscience** is the first scientifically detailed description of developments that will revolutionize most of the industrial processes and products currently in use. This ground breaking work draws on physics and chemistry to establish basic concepts and analytical tools and thus only it provides an indispensable introduction to the emerging field of Nanoscience, Scientific research is the ultimate tool in pushing forward the limit of understanding. But, as with any tool, research is only powerful if used properly, and to its full effects. The more we learn, the more we discover connections threading through biochemical and biophysical world. In writing this Encyclopaedia the author has made every effort to present these connections in a way that will help first time students of Nanoscience understand the subject and how very relevant it is to their lives.

In the preparation of this book large number of books and research papers have been consulted. So no authenticity is claimed.

The author expresses his gratitude to Mr. Wasan and staff of M/s Discovery Publishing House for their whole hearted co-operation in the publication of this book.

The author tried hard to be accurate and upto date in statement and realises the impossibility of completely avoiding errors therefore, the author will greatly appreciate having his attention called to any questionable statement.

Author

Contents

1

INTRODUCTION

Somewhere in your house there is a *refrigerator*, and within it stands a glass bottle of a beverage you wish to drink. Without getting up, or putting down this book, beckon with your hand. The door of the refrigerator will open, and the bottle will emerge, floating in the air. Simultaneously, a crystal goblet will appear in your hand. As the bottle approaches, the stopper will remove itself with a minimum of fuss. Enjoy your refreshment while the bottle floats serenely back to its place.

You decide to relandscape your yard. Go outside: somehow you become 50 feet tall, and the area looks like a model of itself. To your tools, the ground seems to have the consistency of whipped cream; but no matter how careless your step, your feet and knees aren't able to damage anything. After you've swapped the pond and the tennis court, utter a few odd-sounding words. You're normal size, the better to judge the effect. A few more magic words, like "*plinth*" and "*colonnade*," and a small Grecian temple—marble columns and all—appears. Wave your arms and it expands and contracts like some outsized accordion. When it's just right, lock it into place with a final command. Any sufficiently advanced technology, wrote Arthur Clarke, is *indistinguishable* from *magic*. The particular invention is about, "*utility fog*," brings this point home poignantly. *Magic* is what you get when the rules of the mind describe the real, physical world. Objects appear and disappear, like thoughts of. Similarity and contagion—the "*laws*" of magic—are really associative retrieval and processing. With utility fog, the physical works is mind, in an objective literal sense that will become clear in the next few pages.

MICROSCOPIC ROBOT

Imaging a *microscopic robot*, about the size of a single human cell. It has a body about the size of a bacterium and several arms sticking out in all directions. A bucketful of such robots might from a "*robot crystal*" by linking their arms together into a lattice structure. Now take a room, with people, furniture, and other objects in it—it's still mostly empty air—and fill it completely full of robots.

With the right programming, the *robots* can exert any force in any direction on the surface of any object. They can hold an object up, so that it apparently floats in the air. They can hold you up, applying the same pressure to the seat of your pants that your chair would. They can exert the same resisting forces that your elbows and fingertips would expect from the arms and back of the chair. A program running in utility fog can thus simulate the physical existence of almost any object.

Utility fog operates in two modes: first, the "*naive*" mode where the robots act much like cells, and each robot occupies a particular position and does a particular function in a given object. In the second, or "*fog*" mode, the robots act more like the pixels on a TV screen. Objects are then formed of *patterns* of robots, which vary their properties according to which part of the object they are representing at the time. An object can then move through a cloud of robots without moving the individual robots, just as the pixels on your TV stay put while pictures move around on the screen.

Although each of the robots is too small to see individually, masses of them would scatter light so much as to be opaque (just like fog). The robots at the surface of a naive-mode object could hold a layer of surface plates of any desired colour. Unused plates could be distributed throughout the volume of the object an physically swapped into place about as fast as a television changes its picture. In fog mode, the object would have to hold up holographic "*eye-phones*," in front of your eyes, similar to those described elsewhere in this volume.

Utility fog that is pretending to be air needs to be impalpable. You would like to be able to walk through your fog-filled house without having the feeling you had been cast into a solid block of Lucite. You'd also like to be able to breathe. To this end, the robots representing empty space constantly run a fluid-flow simulation of what the air would be doing if the robots weren't there. Each robot does what the air it displaced would do in its absence. How do you breathe when the air is a solid mass of machines? Actually, it isn't solid—

foglets only take up about 5 percent of the volume they occupy (they need lots of "*elbow room*" to move around). Like cotton candy, they carry the air in a given volume along when the fog moves, but relatively little force is required to compress the air out when needed. Thus when fog simulates the normal air currents caused by breathing and other motions, air is moved and mixed just as it normally would be; when you inhale, fog squeezes out the appropriate amount of air.

Carried along in the fog, every six inches or so, is a pea-sized object that acts as a control center and local *power reservoir*. In the case of catastrophic system failure, all the fog that is part of a naive mode object simply freezes into place, and that which is pretending to be air contracts around its local centers. In a typical room, the floor would be covered to a depth of half a foot in objects very similar to golf balls.

To understand why you'd want to fill the air with microscopic robots only to go to so much trouble to make it seem like they weren't there, consider the advantages of a TV or computer screen over an ordinary photograph. Objects on the screen can appear and disappear at will; they are not constrained by the laws of physics. The whole scene can shift instantly from one apparent locale to another. Completely imaginary construction's impossible to build in reality, can be common place. Virtually anything you can imagine could be given tangible reality in a utility fog environment.

Why not, instead, build a virtual-reality machine that fools all your senses into a purely simulated version of the same apparent world? Well, in a *VR machine* you could experience digging up your backyard as a fifty-foot giant, but the backyard itself would miss all the fun. Fog acts as a continuous bridge between actual *physical reality* and *virtual reality*. Fog is a universal effector as well as a universal sensor. Any (real) object in a fog environment can be manipulated with an extremely wide array of pattern of pressure, force, and support; illuminated with any pattern of light; measured, analyzed, smelled, tasted, weighed, cut, reassembled, or reduced to bacteria-sized pieces and sorted for recycling.

Roads Must Roll

As well as forming an extension of the senses and muscles of individual people, *fog* can act as a generalized infrastructure for society at large. *Fog City* need have no permanent building of concrete , no roads of asphalt, no cars, trucks, or buses. It can look like a park, or a forest, or if the population is sufficiently whimsical, *ancient Rome* one day and the *Emerald City* the next. It will be more efficient to build dedicated machine for long distance energy and information propagation and physical transport.

Fog is ideal for local use and as an interface to worldwide communication networks. It can act as shelter, clothing, telephone, computer, and automobile. It can be almost any common household object, appearing from nowhere when you need it (and disappearing when you don't). It gains a certain efficiency from this extreme *polymorphism*; the same fog that was your clothing becomes your bath water and then your bed. Another item of infrastructure that will become increasingly important in the future is information processing. Nanotechnology will allow us toe build some really *monster computers*. Although each foglet will possess a comparatively small processor—which is to say, it will have the power of a current-day supercomputer—there are over *16 billion foglets to a cubic inch*. When those foglets are not doing anything else, that is, when they are simulating the interior of a solid object or air that nothing is passing through at the moment, you could use them to balance your checkbook. Or whatever

What Computers Can't Do

When discussion something as far outside of everyday experience as *utility fog*, it is a good idea to delineate both sides of the boundary. Because fog is capable of so many amazing things, we should point out a few of the things it will be incapable of simulating.

1. Anything requiring really hard surfaces, such as knives, axes, drill bits, nails, and so on. *Individual foglets* are hard, but then so are individual grains of confectioner's sugar. *Naive-mode fog* objects can be about as hard as plastic Fog-mode fog can simulate much harder objects, but only by fooling your senses in a number of ways.
2. Anything requiring high density. *Naive-mode objects* can be about as heavy as wood. *Fog-mode objects* can seem heavier by dynamically supplying the forces you'd feel from weight and inertia.
3. Anything requiring both high strength and low volume. You couldn't make a parachute out of fog. (Unless of course, all the air were filled with it, and then you wouldn't need one.)
4. Anything requiring high heat. A *fog fire* could merrily blaze away on fog logs in your fireplace, and it would feel warm on your skin a few feet away. It would feel just the same if you put your hand into the "*flame*".

5. Anything requiring *molecular manipulation* or *chemical transformation*. *Foglets* are simply on the wrong scale to play with atoms. In particular, they cannot reproduce themselves. On the other hand, they can do things like prepare food the same way a human cook does—by mixing, stirring, and using special-purpose tools designed for them to us.
6. *Fog* cannot simulate food, or anything else destined to be broken down chemically. *Eating fog* would be like eating the same amount of sand or sawdust.

Monsters from the Id

In 1611 William Shakespeare wrote his final play, The Tempest. Three hundred forty-five years later a pair of obscure writers, Irving Block and Allen Adler, updated the *Tempest's* plot into a story called "*Forbidden Planet*" an decorated a modern myth.

"*Forbidden Planet*," more precisely the movie version developed by Cyril Hume, has become a classic cautionary tale for any scenario in which people become too powerful an enormous and powerful machine, capable of projecting objects and forces anywhere in any form, following the mental commands of any Krell. The machine works "*not wisely but too well*," manifesting all the deeply buried subconscious desires of the Krells to destroy each other.

Utility fog will provide humans with powers that approximate those of the fictional Krell machine. Fortunately, we have several centuries of literary tradition to guide as around the pitfalls of hubris made reality. We must study this tradition, or we may be doomed to repeat it—a truth that is by no means limited to utility fog, or indeed to nanotechnology in general.

The first thing we can do is to require fully conscious, unequivocal commands in order for the fog to take any action. Beyond that, we can try to suggest some of the protocols that may be useful in managing the fog in situations where humans are interacting in close physical proximity. Even if we solve the problem of translating one's individual wishes, however expressed, into the quadrillions of sets of instructions sent to individual foglets to accomplish what one desires, the problem of who gets to control which foglets is probably a much more contentious one.

We can, if we desire, physicalize the psychological concept of "*personal space*." Foglets within some distance of each person would be under that person's exclusive control; personal spaces could not merge except by mutual consent. This single protocol could prevent most crimes of violence in our *Hypothetical Fog City*.

A corollary point is that physically perpetrated theft would be impossible in a fog world. It would still be possible by informational means—fraud, cons, hacking, and so on—but fog could be programmed to put ownership on the level of physical law (not that it makes much sense to think of staling a fog-mode object, anyway). Ownership and control of fog would not be greatly different in principle from ownership and control of land. Most of us are used to such laws, but it's actually a complicated bundle of rights: you don't own the road, but you have exclusive rights to the part you're driving on right now. You do own your car, but you will discover limits if your park in a tow-away zone.

Indeed, much of the fog's programming will need to have the character of physical laws. In order for the enormous potential complexity to be comprehensible and thus usable to human beings, it needs to be organized by simple but powerful principles that are consonant with the immense, hard-wired information processing that our sensory systems perform. For example, it would be easy to move furniture (or buildings) by manipulating a scale model with your hands, and easy to see what's going on by looking at the model. However, fog could just as easily have flooded the room with 100 kHz sound, and frequency-scaled the echoes down into your auditory range. A bat would have no trouble with this kind of "*scale model*," but to us it's just noise.

It will be necessary, in general, to arrange the overall control of the fog to be extremely distributed, as local as possible, and *robust* in the presence of failure. When we realize that a single cubic inch of fog represents a computer network of greater complexity and computing capacity than any existing in the world today, the concept of hierarchical control with human oversight can be seen to be hopelessly inadequate. Agoric distributed-control algorithms that barter locally for computational resources offer one possible solution.

With Folded Han s

Another major advantage for *space-filing fog* is safety. In your car (or its nanotech descendant) fog forms a dynamic form-fitting cushion that protects you better than any nylon-fiber seat belt. An appropriately built house filled with fog could even protect its inhabitants from the effects of a *nuclear weapon*—within 95 percent or so of its *lethal blast area*.

There are many more mundane ways for can protect its occupants, not the least being to physically remove bacteria, mites, pollen, an so forth, from the air. A for-filled home would no longer be the place where most accidents happen. First, by performing most household tasks using fog as an instrumentality, the cuts and falls that accompany the use of knives, power tools, ladders, and so forth can be eliminated.

Second, the other major class of *household accidents*, injuries to young children who hurt themselves out of ignorance, can be avoided by a number of means. A child who climbed over a stair rail would float harmlessly to the floor. A child could not pull a bookcase over on himself; falling over would not be among the bookcase's repertoire. Power tools, kitchen implements, and cleaning chemicals would not normally exist; they or their analogues would be called into existence only when needed and then vanish, instead of having to be cleaned and put away.

Outside the home, the possibilities are, if anything, grater. One can easily imagine an "*industrial fog*" factory. It would consist of larger robots, perhaps the size of grains of sand. Of course no one would care that it couldn't simulate the texture an appearance of silk or fine china! Unlike domestic fog, which would have the density and strength of balsa wood, *industrial fog* could have bulk properties resembling hardwood or aluminium. A nanotechnological factory would probably consist of a mass of fog with special-purpose reactors embedded in it, where high-energy chemical transformations could take place. All the physical manipulation, transport, assembly, and so forth would be done by the fog—providing the ultimate in *flexible manufacturing*.

From Earth to Moon

The major components and systems of spaceships will need to be made with special-purpose *nanotechnological mechanisms*, and indeed with such mechanisms pushed much closer to their true capacities than anything we have mentioned heretofore. In the cabin of your spaceship, however, you will want an acceleration couch, a fine application for fog. But when you're not accelerating, which is most of the time, you'd prefer there to be something useful, like empty space.

Put *utility fog* in the cabin and you never have to worry about floating out of reach of a handhold. Instruments, consoles, and cabinets for equipment and supplies are not needed. Items that can not be simulated can be embedded in the fog in what are apparently bulkheads. Fog can add great structural strength to the ship itself; the rest of the structure need be not much more than a balloon. The same is true with your spacesuit: fog inside the suit manages the air pressure and makes motion easy; fog outside gives you extremely fine manipulating ability for various tasks. Of course, like the ship, the suit contains many special-purpose, nonfog mechanisms.

Surrounded your space station with fog; it needs radiation shielding anyway. These would be *big industrial foglets* with lots of redundancy in the mechanism, and even so they may get recycled fairly often. All the stock problems from science fiction movies go away: you never need to go outside to fix something; and if you want to go outside, you can't let go and go spinning off into space. Outside, fog also makes a good tugboat for docking spaceships.

When you homestead your little patch of the Moon, bring along a batch of heavy duty fog as well as special-purpose nanotech power generation and waste recycling equipment. There will be million and one tasks, of the ordinary yet arduous physical kind, that must be done to set up and maintain a self-sufficient household.

Shape of Things to Come

Virtually every physical job humans to could be performed by *utility fog*. Most *nonphysical tasks*—those involving communication and decision making—could be greatly facilitate by *fog-mediated telepresence* and *computational power*.

The appearance and functionality of each individual's environment will be completely programmable, and an extremely wide variety of styles to emerge is expected. However, many of the less lovely aspects of life, such as dirty crowded cities, which are products of necessity, could reasonably be expected to decrease.

To demonstrate how fog would appear to its inhabitants and explain what would actually be happening underneath, let us consider the following vignette. Each statement of a surface effect is followed by a description of how the effect is implemented by the fog.

Fred lives 100 mils away. You subvocalized a command to inquire about his status. Utility for robots in a layer next to your skin, acting as EEG receptors, are able to read enough of your nerve activity to interpret this command. A communication pathway is opened between your domicile and Fred's; this consists of a foglet-to foglet network between you and the nearest hard-matter data transmission node, and along that network to whatever batch of fog Fred has acting as his "social secretary" at the moment, Fred's program knows that he is accepting visitors, at least you at the moment, so an unoccupied section of fog in Fred's place is bound over to your control. A telepresence link is set up; the fog assumes your shape and transmits sights, sounds, and forces to the fog around you, which assumes the appearance of the remote location. To your senses and Fred's, you have been instantly transported to his location.

"*Fred* looked like the very devil, and in fact his place was a pretty decent Hell. Carefully avoiding pit of bubbling lava. I decided the only appropriate shape for me was Frosty the Snowman." Assuming that the appearance and physical properties of a shape are programmed, the software techniques to simulate it are straightforward extension of current-day computer graphics. Fog has more than adequate computational power to do this in real time. Fog around a person can easily simulate an elaborate costume, using, if desired, telepresence techniques to make the person feel as if he actually had taken on that shape.

The zoo is a real place, since people like to see animals in the flesh. You must travel there physically. Each of you will travel in a capsule that simulates the interior of a restaurant, including a telepresence of the other. The restaurant itself is on the other side of the world. In response to your query (and cash!), it downloads the appropriate software for stimulating its interior to both of you.

Food, the most common item fog cannot simulate, is assembled by "*standard*" molecular construction under the direction of other programs downloaded from the restaurant. The developing food forms a tree of coalescing subparts that intercepts your transportation trajectory just as the waiter, who is a program rather than a telepresence, appears to bring it from the kitchen. Even though the substance of fog is not much denser than balsa wood, it can appear to have weight and mass by simulating the forces that gravity and inertia would cause it to exert against your hand. It can do this as long as any part of the object is in contact with the "*air*".

Physicalized "*personal space*" could easily protect people and animals from each other in a zoo, allowing them to intermix freely without cages in perfect safety. For highly nervous animals, it would be possible to program invisibility between them and their predators, or for the visitors if they would overly excite the animals.

Flying individually is not different than using fog to move any other object. It feels, presumably, more or less like swimming. You would almost certainly want to map the control for attitude, sped, and direction into motor impulses instead of verbal commands; it's actually a simple technology, used today for prosthetics.

In the vicinity of the floating city, your trajectory is more closely controlled, so you generally request a least-time routing. The city is a mass of structures of varying permanence; there is no need for streets or walkways, but mall-like promenades are provided in some commercial sectors for that quaint twentieth-century ambiance.

Rossum's Universal Robots

Compared with an assembler, a foglet will be huge and overpowered, able to control its motions to a tenth of a micron instead of a tenth of a nanometer. It will have an arm-spread of ten microns. Each foglet will weigh about 20 nanograms and contain about 5 trillion atoms.

This is about as small as it would be feasible to make a foglet. There is no obvious upper bound on size, except to reduce the resolution and verisimilitude of the simulation. Foglets whose appearance was unimportant and that were simply to manipulate objects could be on the order of inches or even feet. It would probably be workable to have foglets ten or even a hundred times as large as the design presented here, which would simplify some of the engineering problems. They would be visible to the naked eye, if you looked closely, but then so are the pixels on your television.

Most currently proposed nanotechnological designs are based on carbon. *Carbon* is a marvelous atom for structural purposes, forming a crystal (*diamond*) that is very stiff and strong. However a fog built a of diamond would have a problem that nanomechanical designs of a more conventional form do not pose: fog has so much surface area exposed to the air that if it were largely diamond, especially on the surface, it would amount to an easily ignitable fuel-air explosive.

Therefore, the basic *foglet* is designed sot hat its structural elements, forming the major component of its mass, are made of aluminum oxide, a refractory compound using common elements. The structural elements form an exoskeleton, which, besides being a good mechanical design, allows us to have an evacuated interior in which more sensitive nanomechanical components can operate. Of course, any macroscopic ignition source would vapourize a foglet entirely; but as long as more energy is used vapourizing the exoskeleton than is gained burning the carbon-based components inside, the reaction can not spread.

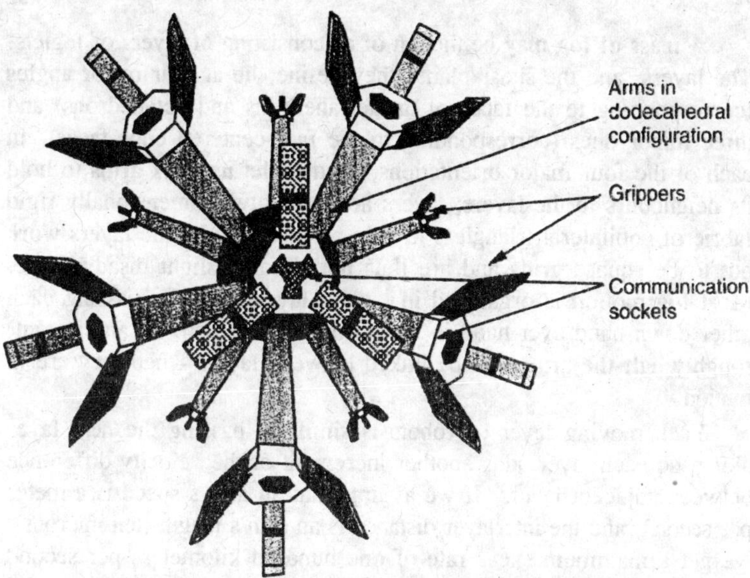

Fig. 1.1. A foglet, containing some five trillion atoms, with a reach of about ten microns.

Each *foglet* has twelve arms, arranged as the faces of a dodecahedron. The arm have telescoping mechanisms rather than joints. Each arm swivels on a universal joint at the base, and the gripper at the end can rotate about the arm's axis. Each arm thus has four degrees of freedom, plus opening and closing the gripper. The swivel and rotate axes are weakly driven, able to position the arm in free air but not drive any kind of load. The only load-carrying motor on each axis is the extension/retraction motor.

The *gripper* is a hexagonal structure with three fingers mounted on alternating faces of the hexagon. Two foglets "clasp hands" in an interleaved six-finger grip. Since the fingers a are designed to match the end of the other arm, this provides a relatively rigid connection; forces are only transmitted axially through the grip.

When at rest, foglets form a regular lattice structure. If the bodies of the foglets are thought of as atoms, it is a *face-centered cubic crystal* formation, in which each atom touches twelve other atoms. If you imagine the foglets' arms as the girders of a trusswork bridge, they form a configuration known as an octet truss, invented by Buckminster Fuller in 1956. The spaces bounded by the arms form alternate tetrahedrons and octahedrons, both of which are rigid, three-dimensional shapes.

A mass of fog may be though of as consisting of layers of foglets. The layers, and the shear planes they define, lie at four major angles (corresponding to the faces of the tetrahedrons and octahedrons) and three minor ones (corresponding to the face-centered cube-faces). In each of the four major orientations, each foglet uses six arms to hold its neighbours in the layers; layers are thus a two-dimensionally rigid fabric of equilateral triangles. In face-centered mode, the layers work out to be square grids and are thus not rigid, a slight disadvantage. Most fog motion is organized in layers; layers slide by passing each other down hand-over-hand in bucket-brigade fashion. At any instant, roughly half the arms will be linked between layers when they are in motion.

Each moving layer of robots is similarly passing the next layer along, so each layer adds another increment of the velocity difference between adjacent layers. If we assume that an arm's speed is a meter per second, and the interlayer distance is an arm's length (ten microns), we get a maximum shear rate of one hundred kilometers per second per meter. This means that if Sandy Koufax were standing in solid fog, pitching a 100 mile-per-hour fastball, the boundary layer of foglets allowing his hand to pass need only be four tenths (0.447) of a millimeter thick.

The atomically precise crystals of the foglet's structural members will have a tensile strength of at least 100,000 pounds per square inch (*psi*), or higher than most steels but lower than modern "*hightech*" composites using refractory ceramics. With an arm's length of ten microns, fog will occupy 5 percent of the volume of the air but has structural efficiency of only about 1 percent in any given direction.

Thus *utility fog*, as a bulk material, will have a density (specific gravity) of 0.2. For comparison, the density of balsa wood is about 0.15, cork is about 0.25. In action, fog will have a tensile strength of only 1000 psi, which is about the same as low-density polyethylene (solid, not foam). Luckily the material properties arising from the lattice structure are more or less isotropic; the one exception is that when fog is flowing, tensile strength perpendicular to the shear plane is cut roughly in half.

Without altering the lattice connectivity, *fog* can contract by up to about 40 percent in any linear dimension, reducing its overall volume (and increasing its density) by a factor of five. (This is of course done by retracting all arms while not letting go.) In this state the fog has the density of water. An even denser state can be attained by forming two interpenetrating lattices and retracting; at this point its density and strength would be similar to ivory or Corian structural plastic: a specific gravity of 2 and tensile strength of about 6000 psi. Such high-density fog would have the useful property of being waterproof (which ordinary fog is not), but it cannot flow and takes much longer to change configuration.

"*Industrial*" fog could, by use of thicker arms, increase the density and strength of ordinary fog by a factor of three, without losing mobility. In a retracted state, properly designed industrial fog could be waterproof without interpenetrating, making it more flexible in applications demanding that property.

Mightiest Machine

For a mass of fog to press against some object at its peak 1000 psi, each foglet arm must sustain a force of about 30 dynes. For the arm to go from fully retracted to fully extended will take 12 milliergs of energy. If the foglet ran on hydrogen, it would burn about 3 billion atoms to get this much energy. That may sound like a lot, but it's less than 0.1 percent of the volume of a foglet, a pretty small fuel tank.

In fact, foglets run on electricity, but they store hydrogen as an energy buffer. Hydrogen is used in part because it's almost certain to be a fuel of choice in the nanotech world, and thus we can be sure that the process of converting hydrogen and oxygen to water and energy—as well as the process of converting energy and water to hydrogen and oxygen—will be well understood. That means we'll be able to make these conversions efficiently, which is of prime importance.

Suppose that the fog is flowing, layers sliding against each other, and some force is being transmitted through the flow. This would happen any time fog moved a nonfog object, for example. Just as your muscles oppose each other when you hold something tightly, opposing forces along different foglet arms act to hold the fog's shape and supply the required motion.

When two layers of fog move past each other, the arms between may need to move as many as 100 thousand times per second. If each of those motions were dissipative, and the fog were under full load, it would need to consume 700 kilowatts per cubic centimeter. This is roughly the power dissipation in a 0.45 caliber cartridge the millisecond after the trigger is pulled—which is to say: it just won't do.

But nowhere near this amount of energy is actually being used; pushing arms are supplying this much force, but the arms being pushed are *receiving* almost the same amount, minus the work being done on the object being moved. So if the motors can act as generators when they're being moved. So if the motors can act as generators when they're being pushed, each foglet's energy budget is nearly balanced. Because these are arms instead of wheels, intake and outflow do not match at any given instant, even though they average out over time. Some buffering is needed. Hence the hydrogen. I should hasten to add that almost never would you expect fog to move actively at 1000 psi; even to raise your body, the pressure in the column of fog beneath you is less than one thousandth of that. The 1000 psi is available so that fog can simulate hard objects, where forces can be concentrated into very small areas.

Even so, current exploratory engineering designs for electric motors have power conversion densities up to a billion watts per cubic centimeter and dissipative inefficiencies in the 10 part per million range. This means that if the Empire State Building were being floated around on a column of fog, the fog would dissipate less than a watt per cubic centimeter. Moving fog will dissipate energy by air turbulence and viscous drag. In the large, air will be entrained in the layers of moving fog and forced into *laminar flow*. Energy consumed in this regime may be properly thought of as necessary for the desired motion. As for the waving of the arms between layers, the Reynolds number decreases linearly with the size of the arm. Since the absolute velocity of the arms is low, that is 1 m/s, the Reynold number should be well below the "*lower critical*" value, and the arms should be operating in a perfectly viscous regime with no turbulence. The remaining effect, *viscous drag* (on the waving arms), comes to a few watts per square meter of shear plane per layer.

There will certainly be some waste heat generated by working fog that will need to be dissipated. This and other applications for heat pumps, such as personal heating or cooling (no need to heat the whole building, especially since people prefer different temperatures), can be accomplished simply by running a flow of fog through a pipe-like volume that changes in area, compressing and expanding the entrained air at the appropriate places.

Marooned in Realtime

In the *macroscopic world*, microcomputer-based controllers (e.g., the widely used Intel 8051 series microcontrollers) typically run at a clock speed of about 10 MHz. They emit control signals on the order of 10 kHz (usually less) and control motions in robots that are at most

10 Hz. Thus a complete motion takes one tenth of a second. A million clock cycles per action is not strictly necessary, of course, but it gives us some concept of the action rate we might expect for a given computer clock rate in a digitally controlled nanorobot. Analysis shows that it is possible to build *mechanical* nanocomputers with gigahertz clock rates. From this it is clear that we can build a nanocontroller that can direct a 10-KHz robot. We can do better, however.

Computer architectures have advanced since the really microcontrollers were developed. The 8051 completes a single instruction in 6, 12, or 18 clock cycles; modern RISC (reduced instruction-set computing) architectures execute one instruction each cycle. So far, nobody has bothered to build a RISC microcontroller; they already have more computing power than they need. RISC designs are also efficient when it comes to hardware; one early RISC chip was implemented on a 10,000-gate gate array. This particular design could be translated into a Drexlerian, rod-logic machine using less than one tenth of one percent of a cubic micron.

Each foglet has twelve arms with three axis controls each. In current technology it isn't uncommon to have a processor *per axis*; we could easily fit thirty-six processors into the foglet, but it isn't necessary. The tradeoffs in macroscopic robotics today are such that processors are cheap; in a foglet things are different. Control of a foglet's arms is actually much simpler than control of a macroscopic robot. They can be managed by much simpler controllers that need to execute commands like, "Move to point x at speed by." Using a RISC design allows a single processor to control a 100 kHz arm; using auxiliary controllers will let it control all twelve easily. But there is still a problem.

There is a fundamental *thermodynamic law* that states that every time a bit is erased in a computer, you must dissipate kT ln 2 joules of energy. This includes not only bits explicitly erased in a register but also ordinary logic operations such as a NOR (not-or) gate where two bits go in a and only one comes out. Rod logic (or other nanotechnological computer hardware) designs are quite efficient in that hey get near this physical limitation; but it's better if they don not. Ten-thousand gates of conventional logic operating at a gigahertz (at room temperature) must dissipate 27 nanowatts of heat even if they are perfectly efficient.

Twenty-seven *nanowatts* may not seem like much, but there are a billion foglets in a cubic centimeter. So each cubic centimeter of fog must emit 27 watts of heat—from its computers alone. One cubic meter would put out 27 megawatts; the volume of your whole house doesn't even bear thinking about. In as much as this is physical limit, some radical rethinking is necessary.

Luckily, some radical rethinking has already been done. The estimates just used for irreversible bit operations were derived from the design of a conventional macroscopic processor. In conventional design, the irreversibility of bit operations is not an issue, but we should design nanoprocessors to eliminate such operations as much as possible. Some recent work on reversible computers by Ralph Merkle, head of the *Computational Nanotechnology Project* at Xerox PARC, and I, show that it is possible to build computers not very different in basic instruction-level organization than conventional ones, but which only execute irreversible operations on certain instructions. Such designs should reduce the cost to an average of 10 bits per cycle.

This would bring the power dissipation down to around 30 milliwatts per cubic centimeter, which is roughly the same as the dissipation required from mechanical motion! As long as the computers can go into a standby mode when the fog is standing still, this is quite workable. Concentrations of heavy work—mechanical or computing—would still require cooling circulation to a degree, but, as we have seen, fog is perfectly capable of doing that.

Ophiuchi Hotline

What about all the other computing overhead for fog? Besides the individual control of its robotic self, each foglet will have to run a portion of the overall distributed control and communications algorithms. We can use another clock-speed-to-capability analogy from current computers regarding communications. *Megahertz-speed computers* find themselves well employed managing a handful of megabit data lines. Again we are forced to abandon the engineering tradeoffs of the macroscopic world: routing a message through any given node need theoretically consumer only a handful of thermodynamically irreversible bit operations; typical communications controllers take millions. Special-purpose message routers designed with these facts in mind must be a part of the foglet.

If *fog* were configured as a store-and-forward network, using packets with an average length of 100 bytes and a 1000-instruction overhead, information could move through the fog at 5 meters per second, or 11 miles per hour. You could beat this by writing the information on a paper airplane and throwing it across the room. This approach represents a highly inefficient use of computation even with special-purpose hardware. A more efficient communication protocol will be needed.

The essence of such a *protocol* will be the ability to skip over intervening foglets. There are two ways this might be done. First, long chains of linked arms could be circuit-switched into end-to-end conductivity, forming "*long wires*" through which a form of "*wormhole*" routing could be achieved. Second, if we use larger foglets, were could add optical waveguides that ran down the foglet's arms. Larger foglets would also reduce the number of links required between any two given points.

Cellular automata-based algorithms might provide another source of efficiency that could be used for simulating materials that fog is supposed to be representing. These algorithms can be extremely parsimonious in irreversible operations, and they are ideally suited for the massively parallel, distributed computational base fog provides.

Real of the Robots

The counterintuitive inefficiency in communications is an example, possibly the most extreme one, of a case where macroscopic mechanisms outperform utility fog at a specific task. Other examples will emerge when we consider nanoengineered macroscopic mechanisms.

We could imagine a robot, human-sized, that was formed of collection of nanoengineered parts held together by a mass of industrial utility fog. The parts might include "*bones*", perhaps diamond-fiber composites, having great structural strength, as well as motors, power sources, and so forth. The parts would form a sort of erector set that the surrounding fog could assemble to perform the task at hand. Fog could perform directly all subtasks not requiring the excessive strength, power, and so forth that the special-purpose parts would supply.

A *fog house*, or city, would resemble a *fog robot* in this regard. The roof of a house might well be specially engineered for specific qualities—waterproofness, solar energy collection, resistance to general abuse, and so on—far exceeding that which ordinary general purpose fog would have. (On the other hand, fog could, if desired, have excellent insulating properties.) Of course, the rot need not be one piece—it might be inch-square tiles half in place by supporting for, and thus be quite amenable to incremental repair and replacement, and rearrangement at the owner's whim, and all the other advantages we expect from a fog house. Other major components that would be profitably built from special-purpose nanomaches are *power* and *communications modules*.

Working on more-efficient protocols, as suggested above, *fog* would form an acceptable communications link from a person to some terminal in the same building, but it would be extremely inefficient for long-haul, high-bandwidth connections such as would be required for telepresence. *Power* is also almost certainly the domain of special-purpose mechanisms. Power is also almost certainly the domain of special-purpose mechanisms. Power transmission in the fog is likely to be limited, although for different reasons from data transmission. Nanotechnology will give us an amazing array of power generation and distribution possibilities; fog can use most of them.

The critical heterogeneous component of fog is the *fog-producing machine*. *Foglets* are not self-reproducing; there is no need for them to be, and it would complicate their design enormously to give them fine, atom-manipulating capability. On imagines a fog generator the size of a bread box producing for a house, or building-sized machines filling cities with fog. Fog itself, of course, would convey raw materials back to the machine.

Tunnel in the Sky

Fog moves an object by setting up a seed-shaped zone around it. The *foglets* in the zone move with the object, forming a fairing that makes the motions around it smoother. If the object is moving fast fog in its path will compress to let it go by. The air in a fog matrix does not have time to move, so the motion is fairly efficient. For slower motions, efficiency is not so important, but if we wish to prevent slow moving, high-pressure areas from interfering with other airflow operations, we can enclose the object's zone in a self-contained convection cell that moves foglets from in front of it to behind.

Cold Equations

Utility fog is actually one of the *simpler nanotech devices*. It does not have to live between your cells, like medical nanorobots. It does not have to manipulate single atoms, like assemblers. It is physically large, so we do not have to push any theoretical design limits to get everything inside. The *fog's motors*, *computers*, and *communication systems* are all well within the limits of conservatively applied engineering principles.

It's bit *ironic* that the hardest part of the fog is the part we can do right now: the software. To be lived in, fog needs to be very reliable. Physically, that's not too hard; a foglet that breaks down becomes a tiny speck of dust and can be cleaned out of the way like all the other specks by the remaining fog. Furthermore, any individual foglet that tried to do the wrong thing wouldn't accomplish much. But let a distributed control program get loose and all sorts of mischief could happen.

Of course, this isn't a problem for fog alone. Almost any believable scenario for future technology involves ever more complex software performing ever more important functions. Already banks, phones, air traffic control, and a host of other institutions that our lives depend on are run by complex, real-time, distributed programs. Perhaps the prospect of living physically embedded in utility fog will enhance the perceived need for simplicity, reliability, and predictability, ultimately improving the quality of all our computational system.

2

MINIATURE MOLECULES

Water spews from the Earth as geysers. It circles the globe as clouds and falls from the sky as the gentle rain or in thundering torrents. Frozen, it covers parts of Antarctica to a depth of 3,000 meters (10,000 feet) or more. Vast oceans of it submerge much of the planet. Water is one of the key ingredients that makes life on Earth possible. It makes up as much as 95 percent of the weight of some living things.

Water is a simple substance containing only three atoms-two of hydrogen and one of oxygen. What does the composition of water tell us about its characteristics, and why is water so important to living systems? We can't answer these questions without more information about atoms in general, and about hydrogen and oxygen atoms in particular. But it is not only the constituents of matter that are important. To understand the Behaviour of something as apparently simple as water, we need to know how its constituent atoms are linked together. The same is true of the other small molecules that are essential to living systems.

The first part of this chapter will address the constituents of matter: atoms—their variety, properties, and capacity to combine with other atoms. Then we'll consider how matter changes. In addition to changes in state (solid to liquid to gas), substances undergo changes that transform both their composition and their characteristic properties. When cells use. oxygen to "burn" glucose, the products are water, carbon dioxide, and energy to power life activities. This transformation is similar to what happens when the fuel propane is burned in a stove, a combustion reaction. By studying simple systems such as the combustion of propane, we can understand better what happens in systems as complicated as living cells. The discussion of general chemical principles and their application to small molecules aids our understanding of the large molecules that form the basis for the life of an organism.

Later in this chapter, we return to a consideration of the structure and properties of water and its relationship to acids and bases. We close with a bridge to the next chapter—a consideration of characteristic groups of atoms that contribute specific properties to larger molecules of which they are part.

Atoms: Constituents of Matter

More than a million million (1×10^{12}) atoms could fit in a single layer over the period at the end of this sentence. Each atom consists of a dense, positively charged nucleus, around which one or more negatively charged electrons move. The nucleus contains one or more protons and may contain one or more neutrons. Atoms and their component particles have mass. Mass is a property of all matter. Measuring mass measures the quantity of matter present. The greater the mass, the greater the quantity of matter.

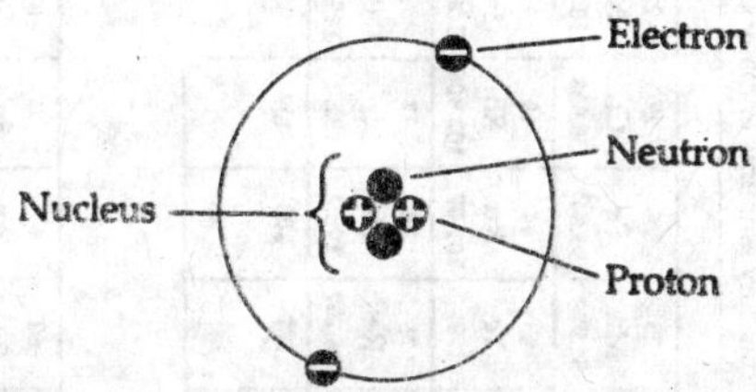

The mass of a proton serves as a standard unit: the atomic mass unit (amu), or dalton (named after the English chemist John Dalton). A single proton or neutron has the mass of 1 dalton, which is 1.7×10^{-24} grams (0.0000000000000000000000017 g). The mass of an electron is 9×10^{-28} g (0.0005 dalton). Because the mass of an electron is so much less than the mass of a proton or a neutron, the contribution of electrons to the mass of an atom can usually be ignored.

The positive electric charge on a proton is defined as a unit of charge. An electron has a charge equal and opposite to that of a proton. Thus the charge of a proton is +1 unit, that of an electron is -1 unit. Unlike charges attract each other; like charges repel. The neutron, as its name suggests, is electrically neutral, so its charge is 0 unit. Because the number of protons in an atom equals the number of electrons, the atom itself is electrically neutral.

An Element is Made Up of Only One Kind of Atom

The element hydrogen consists only of hydrogen atoms; the element iron consists only of iron atoms. An element is a pure substance that

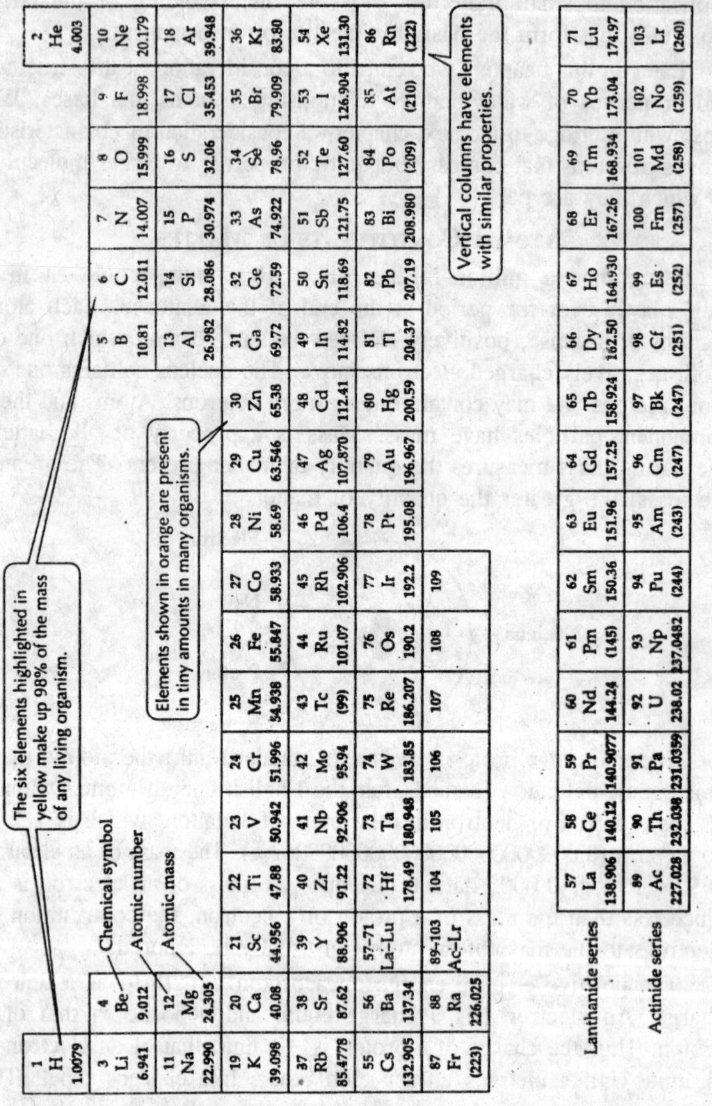

The six elements highlighted in yellow make up 98% of the mass of any living organism.

Elements shown in orange are present in tiny amounts in many organisms.

Vertical columns have elements with similar properties.

(Key: Chemical symbol; Atomic number; Atomic mass)

1 H 1.0079																	2 He 4.003
3 Li 6.941	4 Be 9.012											5 B 10.81	6 C 12.011	7 N 14.007	8 O 15.999	9 F 18.998	10 Ne 20.179
11 Na 22.990	12 Mg 24.305											13 Al 26.982	14 Si 28.086	15 P 30.974	16 S 32.06	17 Cl 35.453	18 Ar 39.948
19 K 39.098	20 Ca 40.08	21 Sc 44.956	22 Ti 47.88	23 V 50.942	24 Cr 51.996	25 Mn 54.938	26 Fe 55.847	27 Co 58.933	28 Ni 58.69	29 Cu 63.546	30 Zn 65.38	31 Ga 69.72	32 Ge 72.59	33 As 74.922	34 Se 78.96	35 Br 79.909	36 Kr 83.80
37 Rb 85.4778	38 Sr 87.62	39 Y 88.906	40 Zr 91.22	41 Nb 92.906	42 Mo 95.94	43 Tc (99)	44 Ru 101.07	45 Rh 102.906	46 Pd 106.4	47 Ag 107.870	48 Cd 112.41	49 In 114.82	50 Sn 118.69	51 Sb 121.75	52 Te 127.60	53 I 126.904	54 Xe 131.30
55 Cs 132.905	56 Ba 137.34	57–71 La–Lu	72 Hf 178.49	73 Ta 180.948	74 W 183.85	75 Re 186.207	76 Os 190.2	77 Ir 192.2	78 Pt 195.08	79 Au 196.967	80 Hg 200.59	81 Tl 204.37	82 Pb 207.19	83 Bi 208.980	84 Po (209)	85 At (210)	86 Rn (222)
87 Fr (223)	88 Ra 226.025	89–103 Ac–Lr	104	105	106	107	108	109									

Lanthanide series	57 La 138.906	58 Ce 140.12	59 Pr 140.9077	60 Nd 144.24	61 Pm (145)	62 Sm 150.36	63 Eu 151.96	64 Gd 157.25	65 Tb 158.924	66 Dy 162.50	67 Ho 164.930	68 Er 167.26	69 Tm 168.934	70 Yb 173.04	71 Lu 174.97
Actinide series	89 Ac 227.028	90 Th 232.038	91 Pa 231.0359	92 U 238.02	93 Np 237.0482	94 Pu (244)	95 Am (243)	96 Cm (247)	97 Bk (247)	98 Cf (251)	99 Es (252)	100 Fm (257)	101 Md (258)	102 No (259)	103 Lr (260)

Fig. 2.1. The periodic table.

contains only one type of atom. The atoms of each element have certain characteristics or properties that distinguish them from the atoms of other elements. The more than 100 elements found in the universe are arranged in the periodic table. The periodic table arranges elements with similar properties in vertical columns in order of their increasing size. Although there are more than 100 elements in the world, about 98 percent of the mass of a living organism (bacterium, turnip, or human) is composed of just six elements—carbon, hydrogen, nitrogen, oxygen, phosphorus, and sulfur. The chemistry of these elements will be our primary concern.

A substance (such as oxygen gas) that contains only one kind of atom is an elemental substance. A substance that contains more than one kind of atom is a *compound*. Most substances of biological interest are compounds.

Number of Protons Identifies the Element

An atom is distinguished from other atoms by the number of its protons, which does not change. This number is called the atomic number. An atom of hydrogen contains 1 proton, a helium atom has 2 protons, carbon has 6 protons, and plutonium has 94. The atomic numbers of these elements are thus 1, 2, 6, and 94, respectively.

Every atom except hydrogen has one or more neutrons in its nucleus. The *mass number* of an atom equals the total number of protons and neutrons in its nucleus. Because the mass of an electron is infinitesimal compared with that of a neutron or proton, electrons are ignored in calculating the mass number. The nucleus of a helium atom contains 2 protons and 2 neutrons; oxygen has 8 protons and 8 neutrons. Helium, therefore, has a mass number of 4 and oxygen a mass number of 16. The mass number may be thought of as the weight of the atom, in daltons.

Each element has its own one- or two-letter symbol. For example, H stands for hydrogen, He for helium, and O for oxygen. Some symbols come from other languages: Fe (from Latin *ferrum*) stands for iron, Na (Latin *natrium*) for sodium, and W (German *Wolfram*) for tungsten. The periodic table gives the symbols for all of the 92 natural elements, as well as those for 14 elements that do not occur naturally.

Isotopes Differ in Number of Neutrons

We have been speaking of hydrogen and oxygen as if each had only one atomic form. But this is not true. Not all atoms of an element have the same mass number. The different atomic forms of a single element are called *isotopes* of the element. Isotopes of an element differ in the number of neutrons in the atomic nucleus. The common form of hydrogen is ^{1}H, but about one out of every 6,500 hydrogen atoms on Earth has a neutron as well as a proton in its nucleus and is thus ^{2}H, called deuterium. Furthermore, it is possible to create ^{3}H, tritium, which has *two* neutrons and a proton in its nucleus. Because all three types of hydrogen atoms have only one proton, they all have the atomic number 1. Deuterium, tritium, and common hydrogen have virtually identical chemical properties, although ^{2}H is twice and ^{3}H three times as heavy as ^{1}H.

In nature, many elements exist as several isotopes. For example, the natural isotopes of carbon are ^{12}C, ^{13}C, and ^{14}C. Unlike the hydrogen isotopes, the isotopes of most other elements do not have distinct names. Rather they are written in the form shown here and are referred to as carbon-12, carbon-13, and carbon-14, respectively. Most carbon atoms are ^{12}C, about 1.1 percent are ^{13}C, and a tiny fraction are ^{14}C. An element's atomic *mass* (*atomic* weight) is the average of the mass numbers of a representative sample of atoms of the element, with all isotopes in their normal proportions. For example, the atomic mass of carbon is 12.011. In biology, one encounters the terms "weight" and "atomic weight" more frequently than "mass" and "atomic mass"; therefore, we will use "weight" for the remainder of this book. Thus we say that the atomic weight of carbon is 12.011.

Some isotopes, called *radioisotopes*, are unstable and spontaneously give off energy as α (alpha), β (beta), or γ (gamma) radiation from the atomic nucleus. Such radioactive decay transforms the original atom into another type, usually of another element. For example, uranium-238 loses an alpha particle to form thorium-234, and carbon-14 loses a beta particle to form nitrogen-14. Biologists can incorporate radioisotopes into molecules and use the emitted radiation as a tag to identify changes that the molecules undergo in the body or to identify the locations of molecules within the cell. Some radioisotopes commonly used in biological experiments are ^{3}H (tritium), ^{14}C (carbon-14), and ^{32}P (phosphorus-32).

Although radioisotopes are useful for experiments and medicine, even low doses of radiation from radioisotopes have the potential to damage molecules and cells. Gamma radiation from cobalt-60 (^{60}Co) is used medically to damage or kill rapidly dividing cancer cells. In addition to these applications, radioisotopes can be used to date fossils.

Electron Behaviour Determines Chemical Bonding

In atoms, biologists are concerned primarily with electrons. To understand organisms, biologists study chemical changes that occur in living cells. These changes, called *chemical reactions* or just *reactions*, are changes in the atomic composition of substances. They occur because of the way in which electrons behave. The characteristic number of electrons in each atom of an element determines how the atom reacts with other atoms. All chemical reactions involve changes in the relationships of electrons with each other.

The location of a given electron in an atom at any given time is impossible to determine. We can only describe a volume of space within the atom where the electron is likely to be. The region of space within which the electron is found at least 90 percent of the time is the electron's orbital. In an atom, a given orbital can be occupied by at most two electrons. Thus any atom larger than helium (atomic number 2) must have electrons in two or more orbitals. The different orbitals have characteristic forms and orientations in space.

The orbitals constitute a series of *electron shells*, or energy levels, around the nucleus. The innermost electron shell, called the *K* shell, consists of only one orbital, called an *s* orbital. The *s* orbital fills first, and its electrons have the lowest energy. Hydrogen ($_1H$) has one *K* shell electron; helium ($_2He$) has two. All other atoms have two *K* shell electrons, as well as electrons in other shells. The *L* shell is made up of four orbitals (an *s* orbital and three *p* orbitals) and hence can hold up to eight electrons. The M, N, O, P, and Q shells have different numbers of orbitals, but the outermost orbitals can usually hold only eight electrons.

In any atom, the outermost shell determines how the atom combines with other atoms; that is, these outermost electrons determine how an atom behaves chemically. When an outermost shell consisting of four orbitals contains eight electrons, the atom is stable and will

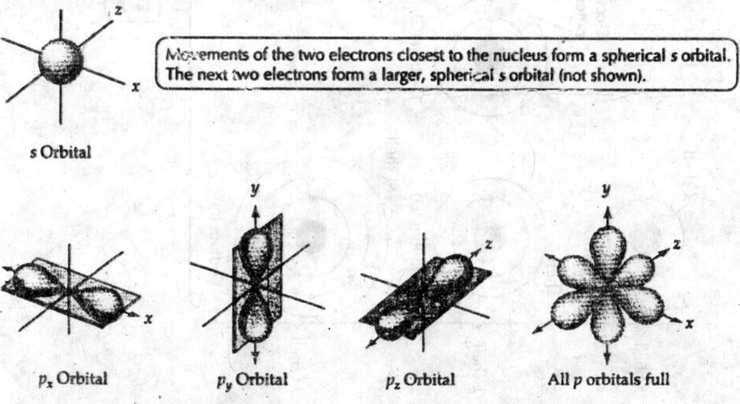

Fig. 2.2. Electron orbitals.

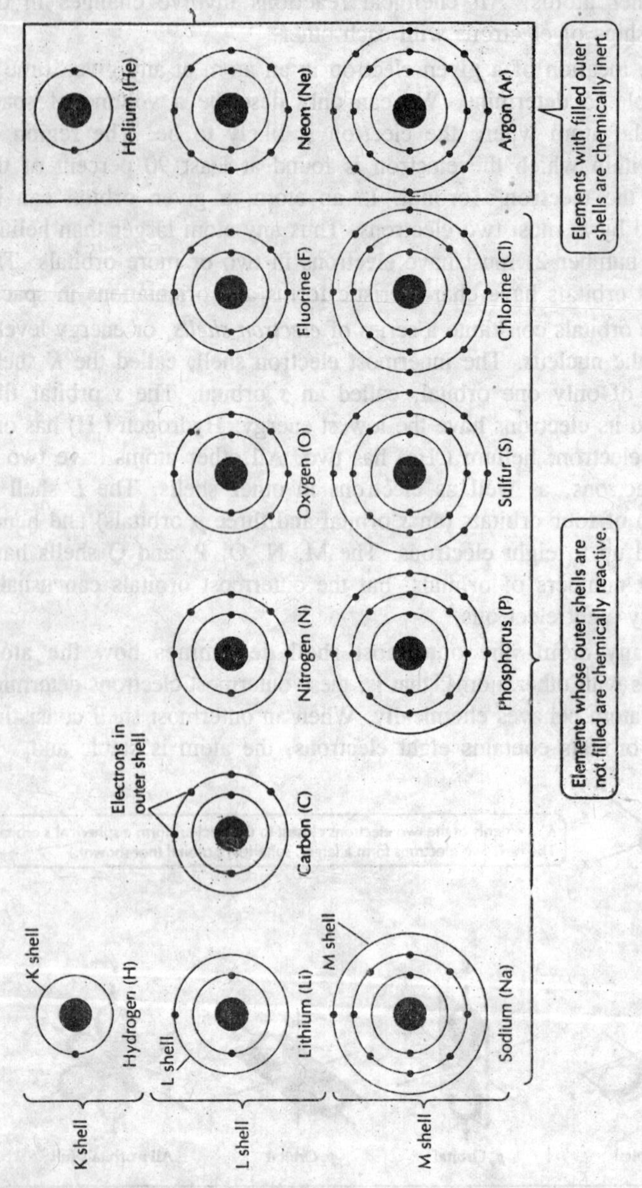

Fig. 2.3. Electron shells determine the reactivity of atoms.

not react with other atoms. Examples of some chemically inert elements are neon and argon, in each of which the outermost shell contains eight electrons. The atoms of other elements seek to attain the stable condition of having filled outer orbitals. They attain this stability by sharing electrons with other atoms or by gaining or losing one or more electrons from their outermost orbitals.

When they share electrons, atoms are bonded together. Such bonds create stable associations of atoms called molecules. A *molecule* can be defined as two or more atoms linked by chemical bonds. The tendency of atoms in stable molecules to have eight electrons in their outermost orbitals is known as the *octet rule*.

Many atoms in biologically important molecules follow the octet rule-for example, carbon (C) and nitrogen (N). However, some biologically important atoms such as hydrogen and phosphorus are exceptions to the rule. Hydrogen (H) attains stability when two electrons occupy its outermost orbital; phosphorus (P) is stable when its outermost orbitals contain ten electrons.

Chemical Bonds: Linking Atoms Together

A *chemical bond* is an attractive force that links two atoms to form a molecule. There are different kinds of chemical bonds, but all strong chemical bonds result from an atom's tendency to attain stability by filling its outermost electron orbitals. Atoms can gain stability in the outermost orbitals by sharing electrons or by losing or gaining one or more electrons. In this section, we will first discuss covalent bonds, the strong bonds that result from sharing of electrons. Then we'll examine hydrogen bonds, which are weaker than covalent bonds but enormously important to biology. Finally, we'll consider ionic bonding, which results when ions form as a consequence of the complete loss or gain of electrons by atoms.

Covalent Bonds Consist of Shared Pairs of Electrons

When two atoms attain stable electron numbers in their outer shells by sharing one or more pairs of electrons, a *covalent bond* forms. A hydrogen atom has one electron in its only shell, but two electrons would be a more stable condition. Imagine two hydrogen atoms, initially far apart but coming closer and closer, until they begin to interact. The negatively charged electron in each hydrogen atom is attracted by the positively charged proton in the nucleus of the other hydrogen atom. When the two atoms are close enough, the two electrons spend time between both nuclei, and the two atoms are covalently bonded together, forming a molecule of hydrogen gas (H_2). The two hydrogen nuclei share the two electrons equally and completely.

The two atoms do not come *too* close together, because their positively charged nuclei strongly repel each other. A certain distance between the coupled atoms gives the most stable arrangement. Pulling the atoms slightly farther apart would require an input of energy because of the "gluing" effect of the shared electrons. Pushing the atoms closer together would require energy because of the mutual repulsion of the protons. So the most stable arrangement of the covalently bonded hydrogens can also be described as an arrangement that has a minimum amount of energy and that is less reactive than are the individual atoms alone, each of which has an incompletely filled orbital in the K shell.

A carbon atom has a total of six electrons; two electrons fill its inner shell and four are in its outer L shell. Because the L shell can hold up to eight electrons, this atom can share electrons with up to four other atoms. Thus it can form four covalent bonds. When an atom of carbon reacts with four hydrogen atoms, a substance called methane (CH_4) forms, resulting from the overlapping of electron orbitals. Thanks to electron sharing, the outer shell of methane's carbon atom is filled with eight electrons, and the outer shell of each hydrogen atom is also filled. Thus four covalent bonds—each bond consisting of a shared pair of electrons—hold methane together.

Orientation of Bonds in Space

Not only the number, but also the spatial orientation, of bonds is important. The four filled orbitals around the carbon nucleus of methane distribute themselves in space so that the bonded hydrogens are directed to the corners of a regular tetrahedron with carbon in the center. Although the orientation of orbitals and shapes of molecules differ depending on the kinds of atoms and how they are linked together, it is essential to remember that all molecules occupy space and have three-dimensional shapes. The shapes of molecules contribute to their biological functions.

Multiple Covalent Bonds

A covalent bond is represented by a line between the chemical symbols for the atoms. Bonds in which a single pair of electrons is shared are called *single bonds* (for example, H-H, C-H). When four electrons (two pairs) are shared, the link is a *double bond* (C=C). In the gas ethylene ($H_2C=CH_2$), two carbon atoms share two pairs of electrons. *Triple bonds* (six shared electrons) are rare, but there is one in nitrogen gas (N-N), the chief component of the air we breathe. In the covalent bonds in these five examples, the electrons are shared more or less equally between the nuclei; consequently all regions of

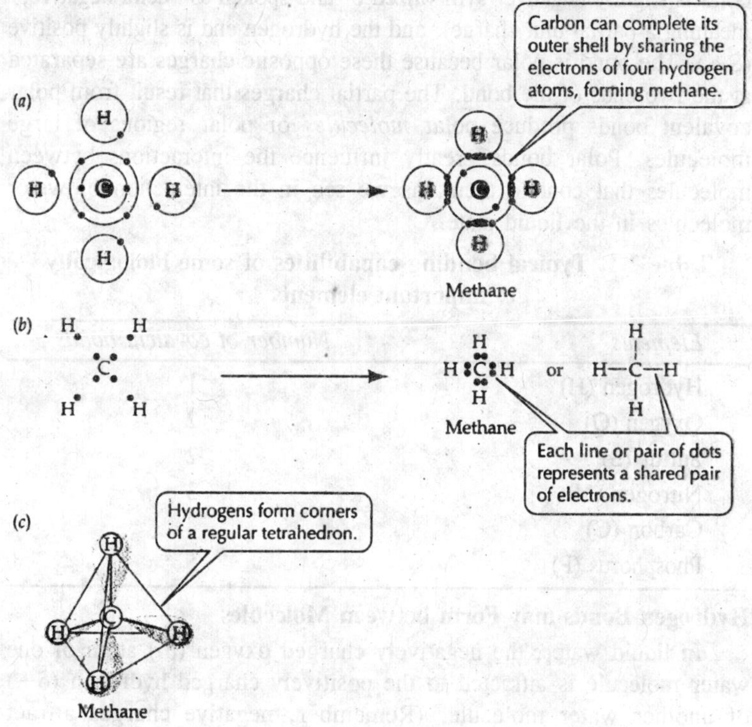

Fig. 2.4. Electrons are shared in covalent bonds.

the bonds are identical. However, when electrons are shared unequally in a covalent bond, regions of partial electric charge exist.

Uniqual Sharing of Electrons

So far we have discussed the covalent bonds that result from the equal sharing of electrons between two nuclei. Now we want to consider the kind of covalent bond that results from unequal sharing of the electrons.

Some atoms hold electrons to themselves more firmly than other atoms do. This characteristic is called *electronegativity*. Highly electronegative atoms that form covalent bonds include oxygen and nitrogen. When these atoms are covalently bonded to atoms with weaker electronegativity, such as carbon and hydrogen, the bonding pair of electrons is unequally shared between the two atoms, and the result is a *polar covalent bond*. For example, when oxygen is bonded to hydrogen, the bonding electrons spend much more time near the oxygen nucleus than near the hydrogen nucleus. Consequently, the oxygen end of the bond is slightly negative (symbolized δ- and spoken as "delta negative," meaning a partial unit charge), and the hydrogen end is slightly positive (S+). The bond is polar because these opposite charges are separated at the two ends of the bond. The partial charges that result from polar covalent bonds produce polar *molecules* or polar regions of large molecules. Polar bonds greatly influence the interactions between molecules that contain them, as we see in the interaction of water molecules in the liquid state.

Table 2.1. Typical bonding capabilities of some biologically important elements

Elements	*Number of covalent bonds*
Hydrogen (H)	1
Oxygen (O)	2
Sulfur (S)	2
Nitrogen (N)	3
Carbon (C)	4
Phosphorus (P)	5

Hydrogen Bonds may Form between Molecules

In liquid water, the negatively charged oxygen (δ-) atom of one water molecule is attracted to the positively charged hydrogen (δ+) of another water molecule. (Remember, negative charges attract positive charges.) The bond resulting from this attraction is called a *hydrogen bond* and is usually symbolized by a series of dots. Hydrogen bonds are not restricted to water molecules. They may form between any covalently bonded hydrogen and an electronegative atom, usually oxygen or nitrogen: -H···O- or -H ···N-. Hydrogen bonds form between small molecules or between different parts of large molecules. Covalent bonds and polar covalent bonds, on the other hand, are always found *within* molecules.

A hydrogen bond is a weak bond; it has about onetwentieth (5 percent) of the strength of a covalent bond between a hydrogen atom and an oxygen atom. However, where many hydrogen bonds form, they have considerable strength and greatly influence the structure and properties of substances. Later in this chapter we'll discuss further how hydrogen bonding in water contributes to many of the properties of water that are significant for living systems. Hydrogen bonds also play important roles in determining and maintaining the three-dimensional shapes of giant molecules such as DNA and protein.

Ions Form Bonds by Electrical Attraction

When one interacting atom is much more electronegative than the other, a complete transfer of one or more electrons may take place. For example, a sodium atom has only one electron in its outermost shell; this condition is unstable. A chlorine atom has seven electrons in its outer shell, another unstable condition. The reaction between sodium and chlorine makes both atoms more stable. When the two atoms meet, the highly *electronegative chlorine atom takes the single unstable* electron from the sodium atom. The result is two electrically charged particles, called *ions*. Ions are electrically charged particles that form when atoms gain or lose one or more electrons.

The sodium ion (Na^+) has a +1 unit charge because it has one less electron than it has protons. The outermost electron shell of the sodium ion is full, with eight electrons, so the ion is stable. The chloride ion (Cl^-) has a –1 unit charge because it has one more electron than it has protons. This additional electron gives Clan outer shell with a stable load of eight electrons. Negatively charged ions are called *anions*; positively charged ions are called *cations*.

Some elements form ions with multiple charges by losing or gaining more than one electron to achieve a stable electron configuration in their outer shell. Examples are Ca^{2+} (the calcium ion, created from a calcium atom that has lost two electrons), Mg^+ (magnesium ion), and Al^{3+} (aluminum ion). Two biologically important elements each yield more than one stable ion: Iron yields Fee^+ (ferrous ion) and Fe^{3+} (ferric ion), and copper yields Cu^+ (cuprous ion) and Cue^+ (cupric ion). Groups of covalently bonded atoms that carry an electric charge are called *complex ions*; examples include NH_4^+ (ammonium ion), SO (sulfate ion), and PO_4^{2-} (phosphate ion).

Once they form, ions 'are usually stable, and no more electrons are lost or gained. As stable entities, ions can enter into stable associations through ionic bonding. Thus stable solids such as sodium chloride (NaCl) and potassium phosphate (K_3P0_4) are formed: Although a very complex solid, bone has as one of its major components the simple ionic compound $Ca_3(PO_4)_2$.

Ionic Bonds: Electrical Attraction

Ionic bonds are the bonds formed by electrical attractions between ions bearing opposite charges. In solids such as table salt (NaCl), the cations and anions are held together by ionic bonds. In solids, the ionic bonds are strong because the ions are close together. However, when ions are dispersed in water, the distance between them can be

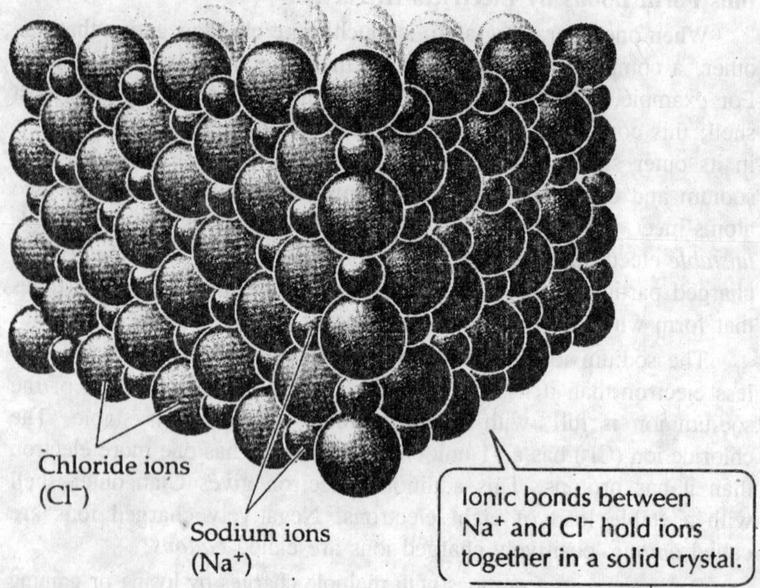

Fig. 2.5. Ionic binding in a solid.

large; the strength of their attraction is thus greatly reduced. Under the conditions that exist in the cell, an ionic bond is less than one-tenth as strong as a covalent bond that shares electrons equally, so an ionic bond can be broken much more readily than a covalent bond.

Not surprisingly, ions with one or more unit charges can interact with polar substances as well as with other ions. Such interaction results when table salt or any other ionic solid dissolves in water. The hydrogen bond that we described earlier is a weak type of ionic bond, because it is formed by electrical attractions. However, it is weaker than most ionic bonds because the hydrogen bond is formed by partial charges (δ- and $\delta+$) rather than by whole unit charges (+1 unit, –1 unit).

Nonpolar Substances have no Attraction for Polar Substances

We have been discussing the bonds that result from electrical attractions between positive and negative charges (ionic bonds and hydrogen bonds). Now let's return to a brief consideration of substances that have "pure" covalent bonds.

These bonds form between atoms that have equal or nearly equal electronegativities-such as carbon and hydrogenwhich share the bonding electrons equally. Such bonds are abundant in the compounds of hydrogen

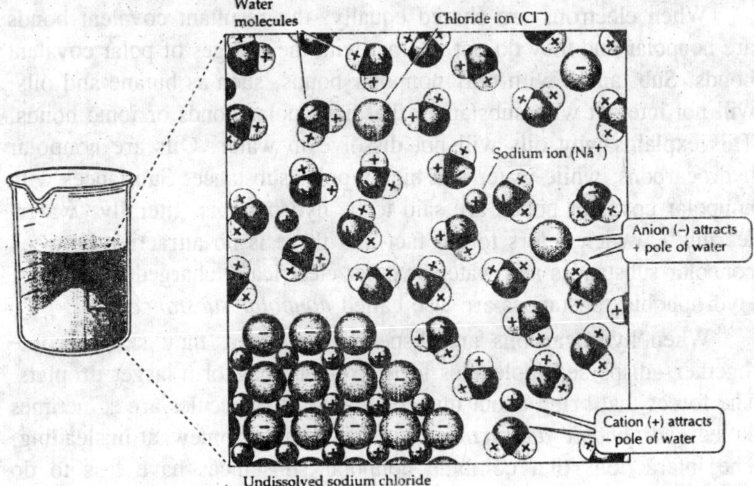

Fig. 2.6. Water molecules surrounded ions.

and carbon—the hydrocarbons. Molecules such as ethane (CH_3—CH_3) and butane (CH_3—CH_2—CH_2—CH_3) are small hydrocarbons, but in living systems, molecules exist with hydrocarbon chains consisting of 16 or more carbon atoms.

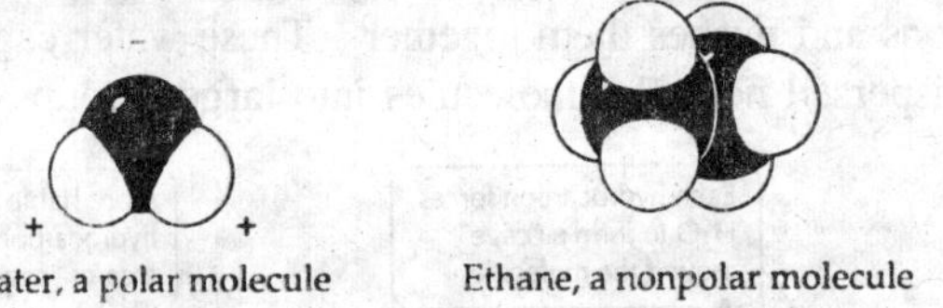

Fig. 2.7. Nonpolar molecules.

Attraction between Nonpolar Molecules

Nonpolar substances such as oils and fats show *van der Waals attractions* between molecules. These attractive forces operate only when nonpolar substances come very close to each other. The random variations in the electron distribution in one molecule create an opposite charge distribution in the adjacent molecule, and the result is a brief, weak attraction. Although each such interaction is brief and weak at any one site, the summation of many such interactions over the entire span of a nonpolar molecule can produce substantial attraction. Thus van der Waals interactions are important in holding together the long hydrocarbon chains that make up the inner portion of biological membranes. They also stabilize portions of the DNA double helix and the intricate folded structure of proteins.

Polar and Nonpolar Interactions

When electrons are shared equally, the resultant covalent bonds are nonpolar and they do not interact with the charges of polar covalent bonds. Substances with only nonpolar bonds, such as butane and oils, will not interact with substances that have polar bonds or ionic bonds. This explains why oils will not dissolve in water. Oils are nonpolar hydrocarbons, while water is a highly polar substance. Substances with nonpolar covalent bonds are said to be hydrophobic (literally "water-fearing"), which refers to the fact that there is no attraction between nonpolar substances and water or other electrically charged substances. Hydrophobic substances are also called *nonpolar substances*.

When hydrocarbons are dispersed in water, they slowly come together—dispersed molecules form droplets that form larger droplets. The forces that bring about this combining of molecules are sometimes called *hydrophobic interactions*, but this term is somewhat misleading. The interactions that combine nonpolar substances have less to do with forces between the nonpolar molecules than with the hydrogen bonding of the water that surrounds the molecules.

When nonpolar substances such as hydrocarbons are introduced into water, they cause a disruption in the usual hydrogen bonding between water molecules. In the vicinity of the hydrocarbon, the water molecules form a hydrogen-bonded "cage" that surrounds the nonpolar hydrocarbons and pushes them together. These water cages can bring together dispersed nonpolar molecules into larger groups.

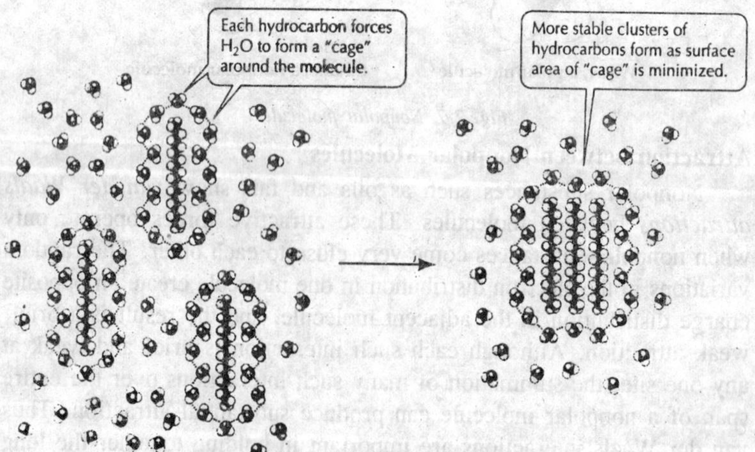

Fig. 2.8. Interaction of water and nonpolar substances.

EGGS BY THE DOZEN, MOLECULES BY THE MOLE

In modern biology, as in chemistry, the question "How much?" is as important as "What kind?" In this section, we will examine briefly how chemists and biologists deal quantitatively with atoms and molecules.

The molecular formula uses chemical symbols to identify different atoms, and subscript numbers to show how many atoms are present. For example, the molecular formula for methane is CH_4 (each molecule contains one carbon atom and four hydrogen atoms), that for oxygen gas is O_2, and that for sucrose (table sugar) is $C_{12}H_{22}O_{11}$. The hormone insulin is represented by the molecular formula $C_{2}5_{4}H_{377}N_{65}O_{76}S_{6}$! Although molecular formulas tell us what kinds of atoms and how many of each kind are present in the molecule, they tell us nothing about which atoms are linked to which. *Structural formulas* give us this information.

Each compound has a *molecular weight* (molecular mass): the sum of the atomic weights of the atoms in the molecule. The atomic weights of hydrogen, carbon, and oxygen are 1.008, 12.011, and 16.000, respectively. Thus the molecular weight of water (H_2O) is $(2 \times 1.008) + 16.000 = 18.016$, or about 18. What is the molecular weight of sucrose ($C_{12}H_{22}O_{11}$)? Your calculations should tell you that the answer is approximately 342. If you remember the molecular weights of a few representative biological compounds, you will be able to picture the relative sizes of molecules that interact with one another. Experiments require quantitative information, Suppose we want to compare how sodium chloride (NaCJ), potassium chloride (KCI), and lithium chloride (LiCI) affect a biological process. At first you might think we could simply give, say, 2 grams (g) of NaCl to one set of subjects, 2 g of KCI to another, and 2 g of LiCI to the third. But because the molecular weights of NaCl, KCI, and LiCI are different, 2-gram samples of each of these substances contain different numbers of molecules. The comparison would thus not be legitimate. Instead, we want to give *equal numbers of molecules* of each substance so that we can compare the activity of one molecule of one substance with that of one molecule of another. How can we measure out equal numbers of molecules?

We can Calculate Numbers of Molecules by Weighing

Measuring the number of molecules is essential to studying how many molecules take part in chemical reactions. Consider two large barrels, one filled with bolts and the other with pennies. How can we determine the number of units in each barrel? We could count them, but that would be tedious (and it is impossible to count individual molecules directly). However, if we know the weight of one bolt and one penny, we can calculate the number of units by dividing the total weight by the weight of each unit. The same principles apply to determining the number of molecules in a weighed sample.

To determine the number of molecules in a sample of a pure substance, we first determine the weight of the substance in grams, then we divide the grams by the relative weight of one molecule (the molecular weight defined earlier). Therefore, to measure out quantities of substances containing equal numbers of molecules, we weigh out in grams a quantity of each substance equivalent to its molecular weight. The molecular weight of NaCl is 58.45. Therefore 58.45 g of NaCl will have the same number of molecules as 74.55 g of KCI, whose molecular weight is 74.55.

If we divide the 74.55 g by the molecular weight of 74.55, the answer is 1. But one what? Surely not one molecule! The calculated numbers here are not the exact number, because we did not divide by the actual weight in grams of a molecule of potassium chloride (which would be about 1×10^{-22} g). Instead we divided by the relative weight. Therefore, instead of knowing the exact number of molecules, we know the relative number, which is called a *mole*. *One mole of a substance is an amount whose weight in grams is numerically equal to the molecular weight of the substance.* Potassium chloride (KCI) has a molecular weight of 74.55, so 1 mole of KCI weighs 74.55 g. Likewise, 1 mole of NaCl weighs 58.45 g, and 1 mole of LiCl, 42.40 g.

A mole of one substance contains the same number of molecules as does a mole of any other substance. This number, known as *Avogadro's number*, is 6.023×10^{23} molecules per mole. The concept of the mole is important for biology because it enables us to work easily with known numbers of molecules. Since we can neither weigh nor count individual molecules, we work with moles. The mole concept is analogous to the concept of a dozen or a gross. We buy some things, such as eggs, by the dozen (or another similar unit, such as the gross) because the individual items are inconvenient to count. Just as a dozen of anything contains twelve of that thing, a mole of any substance contains Avogadro's number of molecules of that substance.

Reactions Take Place in Solutions

In cells and in the laboratory, reactions take place in solutions formed when substances (solutes) are dissolved in water (the solvent).

The amount of substance dissolved in a given amount of water is the concentration of the solution. Knowing the concentrations of solutions is essential in performing experiments and interpreting their results.

A solution that contains 1 mole of solute per liter of solution has a one-molar concentration, abbreviated 1 *M*. A solution containing a half mole per liter is referred to as 0.5 *M*, or half-molar. How would you make 100 milliliters (ml) of a 0.5 *M* sucrose solution? The molecular weight of sucrose is 342, so 1 liter (1,000 ml) of a *1 M* sucrose solution contains 1 mole, or 342 g, of sucrose. You were asked to make just 100 ml of 0.5 *M* solution. Since 34.2 g of sucrose would make 100 ml of *1 M* sucrose, to make 100 ml of a 0.5 *M* sucrose solution, you would use 0.5 × 34.2 g = 17.1 g of sucrose.

Chemical Reactions: Atoms Change Partners

When atoms combine or change bonding partners, a chemical reaction is occurring. Consider the combustion reaction that takes place in the flame of a propane stove. When propane (C_3H_8) reacts with oxygen gas (O_2), the carbon atoms become bonded to oxygen atoms instead of to hydrogen atoms, and the hydrogen atoms become bonded to oxygen instead of to carbon. As the covalently bonded atoms change bonding partners, the composition of the matter changes, and propane and oxygen gas become carbon dioxide and water. This chemical reaction can be represented by the balanced equation

$$C_3H_8 + 5O_2 \rightarrow 3CO_2 + 4H_2O$$

In this equation, the propane and oxygen are the *reactants*, and the carbon dioxide and water are the *products*. The arrow symbolizes the chemical reaction. The numbers preceding the molecular formulas balance the equation and indicate how many molecules react or are produced. In this and all other chemical reactions, matter is neither created or destroyed. The total number of carbons on the left equals the total number on the right. However, there is another product of this reaction: energy.

The heat of the stove's flame and its blue light reveal that the reaction of propane and oxygen releases a great deal of energy. *Energy* is defined as the capacity to do work, but on a more intuitive level, it can be thought of as the capacity to change. Chemical reactions do not create or destroy energy, but *changes* in energy usually accompany chemical reactions. The energy released as heat and light was present in the reactants in another form, called *potential chemical energy*. In some chemical reactions, energy must be supplied from the environment (for example, some substances will react only after being heated), and some of this supplied energy becomes stored as potential chemical energy in the bonds formed in the reactants.

We can measure the energy associated with chemical reactions using the unit called a calorie (cal). A calorie is the amount of heat energy needed to raise the temperature of 1 g of pure water from 14.5°C to 15.5°C. (The nutritionist's Calorie, with a capital C, is what biologists call a kilocalorie (kcal) and is equal to 1,000 heat-energy calories.) Another unit of energy that is increasingly used is the joule (J). When you compare data on energy, always compare joules to joules and calories to calories. The two units can be interconverted: 1 J = 0.239 cal, and 1 cal = 4.184 J. Thus, for example, 486 cal = 2033 J, or 2.033 kJ. Although defined in terms of heat, the calorie and the joule are measures of any form of energy-mechanical, electric, or chemical.

Within living cells, chemical reactions called oxidation-reduction reactions take place that have much in common with this combustion of propane. The fuel for these biological reactions is different (the sugar glucose, rather than propane), and the reactions proceed by many intermediate steps that permit the energy released from the glucose to be harvested and put to use by the cell. But the products are the same: carbon dioxide and water. We will present and discuss oxidation-reduction reactions and several other types of chemical reactions that are prevalent in living systems in the chapters that follow.

WATER: STRUCTURE AND PROPERTIES

Water, like all other matter, can exist in three states: solid (ice), liquid, and gas (vapour). Liquid water is the medium in which life originated on Earth more than 3.5 billion years ago, and it is in water that life evolved for about a billion years. Today water covers three-fourths of Earth's surface, and all active organisms contain between 45 and 95 percent water. No organism can remain biologically active without water, which is important in both the outer and inner environments. Within cells, water participates directly in many chemical reactions, and it is the medium (or solvent) in which most reactions take place. In this section we will consider the structure and interactions of water molecules, exploring how these generate properties essential to life.

Water has a Unique Structure and Special Properties

The shape of a water molecule is determined by the distribution in space of the four pairs of electrons in the outer shell of the oxygen atom. Each pair of electrons is confined to an orbital. Two of these pairs are bonded to hydrogens, but the other two pairs (nonbonding pairs) also influence shape. Because of their negatively charged electrons, the four orbitals, each containing a pair of electrons, repel one another. In seeking to be as far apart as possible, they give the water molecule a nearly tetrahedral shape.

The shape of the water molecule, its polar nature, and the formation of hydrogen bonds between water molecules or with other substances give water its unusual properties. For example, ice floats, and compared to other liquids, water is an excellent solvent, making it an ideal medium for biochemical reactions. Water is both cohesive (sticking to itself) and adhesive (sticking to other things). And the energy changes that

Ice Floats

In its solid state (ice), water is held by its hydrogen bonds in a rigid, crystalline structure in which each water molecule is hydrogen-bonded to four other molecules. Although these molecules are held firmly in place, they are not as tightly packed as they are in liquid water. In other words, *solid water is less dense than liquid water*, which is why ice floats in water. If ice sank in water, as almost all other solids do in their corresponding liquids, ponds and lakes would freeze from the bottom up, becoming solid blocks of ice in winter and killing most of the organisms living in them. Once the whole pond had frozen, its temperature could drop well below the freezing point of water. However, because ice floats, it forms a protective insulating layer on the top of the pond, reducing heat flow to the cold air above. Thus fish, plants, and other organisms in the pond can survive the winter at temperatures no lower than 0°C, the freezing point of pure water.

Melting and Freezing

Compared to other nonmetallic substances of the same size, ice requires a great deal of heat energy to melt. Melting 1 mole of water molecules requires the addition of 5.9 kJ of energy. This value is high because more than a mole of hydrogen bonds must be broken for 1 mole of water to change from solid to liquid. In the opposite process, freezing, a great deal of energy must be lost for water to transform from liquid to solid. These properties help make water a moderator of temperature changes.

Another property of water that moderates temperature is the high heat capacity of liquid water. *Heat capacity* is the amount of heat energy that is required to raise the temperature of a substance 1°C.

Raising the temperature of liquid water takes a relatively large amount of heat. The temperature of a given quantity of water is raised only 1°C by an amount of heat that would increase the temperature of the same quantity of ethyl alcohol by 2°C, or of chloroform by 4°C. This phenomenon contributes to the surprising constancy of the temperature of the oceans and other large bodies of water through the seasons of the year. The temperature changes in coastal land masses are also moderated by large bodies of water. Indeed, water helps minimize variations in atmospheric temperature throughout the planet.

Cohesion and Surface Tension

In liquid water, the molecules are free to move about. The hydrogen bonds between the water molecules continually form and break. In other words, liquid water has a dynamic structure. On the average, every water molecule forms 3.4 hydrogen bonds with other water molecules. This number represents fewer bonds than exist in ice, but it is still a high number.

These hydrogen bonds explain the cohesive strength of water. The *cohesive strength* of water is what permits narrow columns of water to stretch from the roots to the leaves of trees more than 100 meters high. When water evaporates from leaves, the entire column moves upward in response to the pull of the molecules at the top.

Water also has a high *surface tension*, which means that the surface of water exposed to the air is difficult to puncture. The water molecules in this surface layer are hydrogen-bonded to other water molecules below. The surface tension of water permits a container to be filled slightly above its rim without overflowing, and it permits small animals to walk on the surface of water.

Evaporation and Cooling

Water has a high *heat of vaporization*; which means a lot of heat is required to change water from its liquid state to its gaseous state (the process of evaporation). This heat is absorbed from the environment in contact with the water. Evaporation thus has a cooling effect on the environment-whether a leaf, a forest, or an entire land mass. This effect explains why sweating cools the human body: As the sweat evaporates off the skin, it takes with it some of the adjacent body heat.

Water Molecules Sometimes Form Ions

The water molecule has a slight but significant tendency to come apart into a hydroxide ion (OH^-) and a hydrogen ion (a proton, H^+). Actually, *two* water molecules participate in ionization.

The hydronium ion is in effect a. hydrogen ion bound to a water molecule. For simplicity, biochemists tend to use a modified representation of the ionization of water:

$$H_2O \rightarrow H^+ + OH^-$$

Even though only about one water molecule in 500 million is ionized at any given time, the transformation is significant because H^+ and OH^- ions participate in many important biochemical reactions.

ACIDS, BASES, AND THE pH SCALE

The ionization of water is very important for all living creatures. This fact may seem surprising, since the ionization is so slightonly one ionization out of 5×10^8 water molecules. But we are less surprised if we focus on the abundance of water in living systems and the reactive nature of H^+ produced by that ionization. Remember that H^+ is a proton, a tiny bit of charged matter-smaller than any atom or molecule in the cell.

Usually it attaches to another water molecule, but in the cell it can attach to other molecules and substantially change their properties. Because acids and bases donate and accept H^+, they can profoundly alter the water environment in which the chemical reactions and processes of life take place.

In this section we'll examine acids, bases, and their measurement using the pH scale. We'll close by looking at how buffers limit the changes in pH.

Acids Donate H^+, Bases Accept H^+

In pure water, the concentration of hydrogen ions exactly equals that of hydroxide ions (OH^-), and this "solution" is said to be *neutral*. Now suppose we add some HCl (hydrochloric acid). As it dissolves, the HCl ionizes, releasing H^+ and Cl^- ions:

$$HCl \rightarrow H^+ + Cl^-$$

Now there are more H^+ than OH^- ions. Such a solution is *acidic*. A *basic*, or alkaline, solution is one in which there are more OH^- than H^+ ions. A basic solution can be made from water by adding, for example, sodium hydroxide (NaOH), which ionizes to yield OH^- and Na^+ ions, thus making the concentration of OH^- ions greater than that of H^+ ions.

$$NaOH \rightarrow Na^+ + OH$$

An acid is any compound that can *release* H^+ ions in solution. HCl is an acid, as is H_2SO_4 (sulfuric acid). One molecule of sulfuric acid may ionize to yield two H+ ions and one $S04^-$ ion. Biological compounds such as acetic acid and pyruvic acid, which contain -COOH (the carboxyl group) are also acids, because —COOH → -COO^- + H^+.

Bases are compounds that can *accept* H^+ ions. These include the bicarbonate ion (HCO_3^-), which can accept a H^+ ion and become carbonic acid (H_2CO_3); ammonia (NH_3), which can accept a H+ ion and become an ammonium ion (NH_4^+); and many others.

Note that although —COOH is an acid, —COO^- is a base, because -COO^- + H^+ → —COOH. Acids and bases exist as pairs, such as —COOH and —COO^-, because any acid becomes a base when it releases a proton, and any base becomes an acid when it gains a proton.

You may have noticed that the two reactions just discussed are the opposites of each other. The reaction that yields —COO^- and H^+ is reversible and may be expressed as

$$—COOH \rightleftharpoons —COO^- + H^+$$

A *reversible reaction* is one that can proceed in either direction-left to right or right to left-depending on the relative starting concentrations of reacting substances and products. In principle, all chemical reactions are reversible.

pH is the Measure of Hydrogen Ion Concentration

The terms "acidic" and "basic" refer only to *solutions*. How acidic or basic a solution is depends on the relative concentrations of H^+ and OH^- ions in it. "Acid" and "base" refer to *compounds* and *ions*. A compound or ion that is an acid can donate H^+ ions; one that is a base can accept H^+ ions.

How do we specify how acidic or basic a solution is? First, let's look at the H^+ ion concentrations of a few contrasting solutions. In pure water, the H+ concentration is 10^{-7} *M*. In 1 M hydrochloric acid, the H^+ concentration is 1 *M;* and in 1 *M* sodium hydroxide, the H^+ concentration is 10^{-14} *M*. Because its values range so widely-from more than 1.0 *M* to less than 10^{-14} M the H^+ concentration itself is an inconvenient quantity. It is easier to work with the logarithm of the concentration, because logarithms compress this range. We indicate how acidic or basic a solution is by its pH (a term derived from "potential of Hydrogen"). The pH value is defined as the negative logarithm of the hydrogen ion concentration in moles per liter (molar concentration). In chemical notation, molar concentration is often indicated by putting brackets around the symbol for a substance; thus $[H^+]$ stands for the molar concentration of H^+. The equation for pH is

$$pH = -\log_{10}[H^+]$$

Since the H^+ concentration of pure water is 10^{-1} *M*, its pH is -log(10^{-7}) = -(-7), or 7. A smaller negative logarithm means a larger number. In practical terms, a lower pH means a higher H^+ concentration, or greater acidity. In 1 *M* HCl, the H^+ concentration is 1 *M*, so the pH is the negative logarithm of 1 (-log 10^0), or 0. The pH of 1 *M* NaOH is the negative logarithm of 10^{-14}, or 14. A solution with a pH of less than 7 is acidic: It contains more H^+ ions than OH ions. A solution with a pH of 7 is neutral. And a solution with a pH value greater than 7 is basic. Because the pH scale is logarithmic, the values are exponential: A solution with a pH of 5 is 10 times more acidic than one with a pH of 6 (it has ten times as great a concentration of H^+); a solution with a pH of 4 is 100 times more acidic than one with a pH of 6.

Buffers Minimize pH Changes

An organism must control the chemistry of its cellsin particular, the pH of the separate compartments within cells. Animals must also control the pH of their blood. The normal pH of human blood is 7.4, and deviations of even a few tenths of a pH unit can be fatal. The control of pH is made possible in part by *buffers*, systems that maintain a relatively constant pH even when substantial amounts of acid or base are added. A buffer is a mixture of an acid that does not ionize completely in water and its corresponding base-for example, carbonic acid (H_2CO_3) and bicarbonate ions (HCO_3^-). If acid is added to this buffer, not all the H^+ ions from the acid stay in solution. Instead, many of the added H^+ ions combine with bicarbonate ions to produce more carbonic acid, thus using up some of the H+ ions in the solution and decreasing the acidifying effect of the added acid:

$$HCO_3 + H^+ \rightleftharpoons H_2CO_3$$

If base is added, the reaction reverses. Some of the carbonic acid ionizes to produce bicarbonate ions and more H^+, which counteracts some of the added base.

In this way, the buffer minimizes the effects of added acid or base on pH. A given amount of acid or base causes a smaller change in pH in a buffered solution than in an unbuffered one. Buffers illustrate the reversibility of chemical reactions: Addition of acid drives the reaction in one direction; addition of base drives it in the other.

Properties of Molecules

Molecules vary in size. Some are small, such as H_2 and CH_4. Others are larger, such as a molecule of table sugar (sucrose), which has 45 atoms. Still other molecules, such as proteins, are *gigantic*, sometimes containing tens of thousands of *atoms bonded* together in specific ways. Whether large, *medium*, or small, most of the molecules in living systems contain carbon atoms and are thus referred to as *organic molecules*. Most organic molecules include hydrogen and oxygen atoms, and many also include nitrogen and phosphorus.

All molecules have a specific three-dimensional shape. For example, the orientation of the bonding orbitals around the carbon atom gives the molecule of methane (CH_4) the shape of a regular tetrahedron, while in carbon dioxide (CO_2) the three atoms are in line with one another. Larger molecules have specific complex shapes that result from the number and kinds of atoms present and the ways in which these atoms are linked together. Some large molecules have compact ball-like shapes. Others are long, thin, ropelike structures. The shapes relate to the roles these molecules play in living cells.

In addition to size and shape, molecules have certain properties that characterize them and determine their biological roles. Chemists can use the characteristics of composition, structure (three-dimensional shape), reactivity, and solubility to distinguish a sample of one pure compound from another. For example, a compound such as sugar is soluble in water, which means that the solid disperses to form a uniform homogeneous mixture, but the same compound is insoluble in oil (the solid sugar remains a solid), and the resulting mixture is heterogeneous. Another substance, such as butane, is insoluble in water but highly soluble in oil. These solubility differences are due to the different atoms present and to their arrangement in these two kinds of molecules.

The presence of polar or charged sites on a molecule plays important roles in determining the molecule's solubility in water. Such sites can also determine the kinds of chemical reactions in which the molecule participates. The sizes, shapes, solubilities, and reactivities of molecules are significant to understanding the structures and operations of cells and organisms. Certain groups of atoms found together in a variety of molecules simplify our understanding of the reactions that molecules undergo.

Functional Groups give Specific Properties to Molecules

On the basis of atomic composition, structure, and reactivity, we can distinguish families of molecules from other families. Each member within a family of compounds has some characteristics in common with the other members of that family but differs from them in other characteristics. For example, organic acids are a family of carbon compounds that are all acidic but differ in other ways. They all contain a characteristic group of atoms called the *carboxyl group*,

$$\begin{array}{c} O \\ || \\ -C-OH \end{array}\text{, or COOH}$$

that is the source of the H^+ that defines an acid:

$$\begin{array}{c} O \\ || \\ -C-OH \end{array} \rightarrow \begin{array}{c} O \\ || \\ -C-O^- \end{array} + H^+$$

The carboxyl group is a functional group. Functional groups are groups of atoms that are part of a larger molecule and have particular reactive characteristics. The same functional group may be part of very different molecules. In addition to the carboxyl group, you will encounter several other functional groups in your study of biology.

Several classes of biologically important compounds are defined by the functional groups they contain. When the functional group is a hydroxyl group (—OH), the product is an alcohol. Perhaps the most familiar alcohol is ethanol (also called ethyl alcohol, CH_3CH_2OH). Small alcohols like ethanol are soluble in water because of the hydrogen bonding possible between the polar hydroxyl group and water molecules, but larger alcohols are not soluble in water because of their long hydrocarbon chains.

Sugars contain both hydroxyl and carbonyl groups. The carbonyl *group* has a central carbon atom with a double bond to an oxygen atom (C=O). If one of the other two bonds of the carbon atom in a carbonyl group is attached to a hydrogen atom, the compound is an *aldehyde*.

Carboxyl groups are found in organic acids. Some organic bases, called amines, possess an *amino group* ($—NH_2$), which has a tendency to react with H^+ to produce the positively charged ammonium group ($—NH_3^+$). This H^+-accepting characteristic accounts for the classification of amines as bases.

Amino acids are important compounds that possess both a carboxyl group and an amino group attached to the same carbon atom, the a (alpha) carbon. Also attached to the a carbon atom are a hydrogen atom and a side chain designated by the letter R. Different side chains have different chemical compositions, structures, and properties. Each of the 20 amino acids found in proteins has a different side chain that gives it its distinctive chemical properties. Because they possess both carboxyl and amino groups, amino acids are simultaneously acids and bases. At the pH values commonly found in cells, both the carboxyl and the amino groups are ionized: The carboxyl group has lost α proton, and the amino group has gained one.

Two other functional groups deserve our attention: the sulfhydryl group and the phosphate group. The *sulfhydryl group* (—SH) is an important constituent of the side chains of two amino acids. When two of these sulfhydryl groups react, the two hydrogen atoms are lost and a covalent bond, called a disulfide bond, is formed between the sulfur atoms (—S—S—). As we'll see in the next chapter, disulfide bonds help maintain the three-dimensional structure of proteins that is required for normal protein functioning.

The *phosphate group* ($—OPO_3^-$) is found on many different kinds of molecules. It plays important structural roles in DNA and RNA and participates in reactions that transfer energy. The removal (hydrolysis) of phosphate groups from some molecules releases energy that can be used to fuel other energy-requiring reactions.

Isomers have Different Arrangements of the Same Atoms

Isomers are compounds that have the same chemical formula but different arrangements of the atoms. (The prefix "iso-" means "same and is encountered in many technical terms.) Of the different kinds of isomers, we will consider two: structural isomers and optical isomers.

Structural isomers are isomers that differ in how the atoms are joined together. Consider two simple molecules, each composed of four carbon and ten hydrogen atoms bonded covalently, with the formula C_4H_{10}. These atoms can be linked in two alternative ways in which carbon can form four bonds and hydrogen one bond. These two forms are called butane and isobutane. Their different bonding relationships are distinguished in structural formulas, and they have different chemical properties.

$$\begin{array}{ccccccc} & & H & & H & & \\ & & | & & | & & \\ H_3C & — & C & — & C & — & CH_3 \\ & & | & & | & & \\ & & H & & H & & \end{array} \qquad\qquad \begin{array}{ccccc} & & CH_3 & & \\ & & | & & \\ H_3C & — & C & — & CH_3 \\ & & | & & \\ & & H & & \end{array}$$

Butane　　　　Isobutane

Many molecules of biological importance, particularly the sugars and amino acids, have optical *isomers*. Optical isomers (also called enantiomers) are related to each other in the way an object is related to its mirror image. Optical isomers occur whenever a carbon atom has four *different* atoms or groups attached to it. These instances allow two different ways of making the attachments, each the mirror image of the other. Such a carbon atom is an asymmetric carbon, and the pair of compounds are optical isomers of each other. Your right and left hands are optical isomers. Just as a glove is specific for a particular hand, so some biochemical molecules can interact with a specific optical isomer of a compound but are unable to "fit" the other.

The a carbon in an amino acid is an asymmetric carbon because it is bonded to four different groups. Therefore, amino acids exist in two isomeric forms, called D-amino acids and 1.-amino acids. "D" and "L" are abbreviations for right and left, respectively. Only L-amino acids are commonly found in most proteins of living things. The compounds discussed in this chapter include some of the more common ones found in organisms. Between these small molecules and the world of the living stands another level, that of the giant macromolecules. These huge molecules—the proteins, lipids, carbohydrates, and nucleic acids—are the subject of the next chapter.

3

GIANT MOLECULES

The lives of cells are dances and dramas with tens of thousands of different kinds of molecules. Their dramatic choreography is what scientists reveal as they investigate the molecules of living systems, their physical properties, and their chemical reactions. What molecules are present? What are their chemical structures and properties? And what biological functions do these molecules perform? In different ways, we will be concerned with these questions throughout much of this book.

In this chapter, we'll look at gigantic molecules called macromolecules (*macro-*, "large") and the subunits from which they are constructed. Macromolecules may contain hundreds or thousands of atoms and have very large molecular weights. For example, human hemoglobin has a molecular weight of 64,500 and contains more than 6,000 atoms. The molecular complexity of such a structure can be understood in terms of the structures and properties of its subunits, which are fewer in number and easier to understand. Macromolecules are chains of small, individual units called monomers (*mono-*, "one"; *-mer*, "unit") covalently bonded together to form a polymer (*poly-*, "many").

In living systems there are three types of macromolecules: polysaccharides, proteins, and nucleic acids. *Polysaccharides* are constructed from sugar monomers such as glucose; *proteins* are constructed from amino acids; and *nucleic acids* are composed of nucleotides. Another important group of molecules that can form large structures is the *lipids*. But because the individual components (the lipids) do not covalently bond together, the resulting large structures are not true macromolecules.

The monomers that form the basis of polysaccharides, proteins, and nucleic acids are identical in different species. For example, a molecule of glucose in human blood is identical to a molecule of glucose from a reptile or a cabbage plant. The same group of 20 different amino acids is used to construct proteins for all living things—from bacteria to birds, bats, and humans. However, this cannot always be said of the larger molecular forms built from these monomers.

Proteins and nucleic acids in particular are informational macromolecules that are different in different species. For example, human hemoglobin is different from the hemoglobin found in fish or birds. No animal acquires its macromolecules directly from its food. Instead, it uses the subcomponents of its food to construct new macromolecules suited to its unique needs. For example, we eat proteins constructed by other animals and plants, but we break these proteins down into their amino acids and then reassemble the amino acids into the chemically different proteins of our own bodies. This process is like picking up Lego toys that somebody else has made, taking them apart, and putting the parts together again to make the toys we want.

Before we turn to the main focus of this chapter—the macromolecules: carbohydrates, proteins, and nucleic acids—let's first take a look at lipids and the large macromolecule-like structures that they form.

LIPIDS: WATER-INSOLUBLE MOLECULES

Lipids are a chemically diverse group of hydrocarbons. The property they all share is an insolubility in water that is due to the presence of many nonpolar covalent bonds. Nonpolar molecules can associate together and form massive structures, but these structures are not considered macromolecules, because the bonds between the separate molecules are not covalent bonds. These nonpolar hydrocarbon molecules are literally pushed together by surrounding water molecules, which are not attracted to the nonpolar substance. When the nonpolar molecules are sufficiently close together, weak but additive van der Waals forces hold them together. Although insoluble in water, lipids are soluble in other lipids or in nonpolar solvents such as ether or benzene.

In addition to their diverse chemical structures, lipids have many different biological roles. Some of them store energy (the fats and oils). Others play important structural roles in cell membranes (phospholipids). The carotenoids help plants capture light energy, and

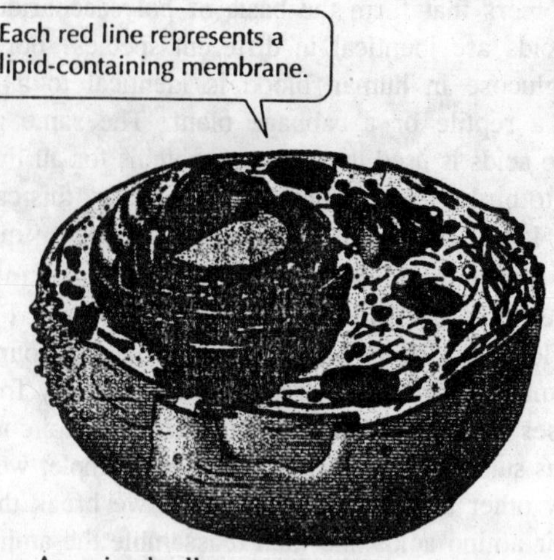

Fig. 3.1. Lipids from cellular membranes.

the steroids and some lipids play regulatory roles. We'll begin our discussion by considering the structures and properties of the fats and oils.

Fats and Oils Store Energy

Chemically, fats and oils are triglycerides, also known as *simple lipids*. Triglycerides that are solid at room temperature (20°C) are called fats; those that are liquid at room temperature are called *oils*.

Triglycerides are composed of two types of building blocks: fatty acids and glycerol. *Glycerol* is a small molecule with three hydroxyl (—OH) groups. Fatty *acids* have long nonpolar hydrocarbon tails and a polar carboxyl functional group (—COOH). Fatty acids can be of different lengths (usually 16 to 20 carbon atoms) and may be either Theturated or unsaturated.

In *saturated fatty acids*, all the bonds between the carbon atoms in the hydrocarbon chain are single bonds-there are no double bonds. That is, all the bonds are *saturated* with hydrogen atoms. Saturated fatty acids include palmitic acid, which has 16 carbon atoms, and stearic acid, which has 18. These molecules are relatively rigid and straight, and they pack together tightly, like pencils in a box.

In *unsaturated fatty acids*, the hydrocarbon chain contains one or more double bonds. Oleic acid is a monounsaturated fatty acid that has 18 carbon atoms, and one double bond near the middle of the hydrocarbon chain causes a kink in the molecule. Fatty acids, such as

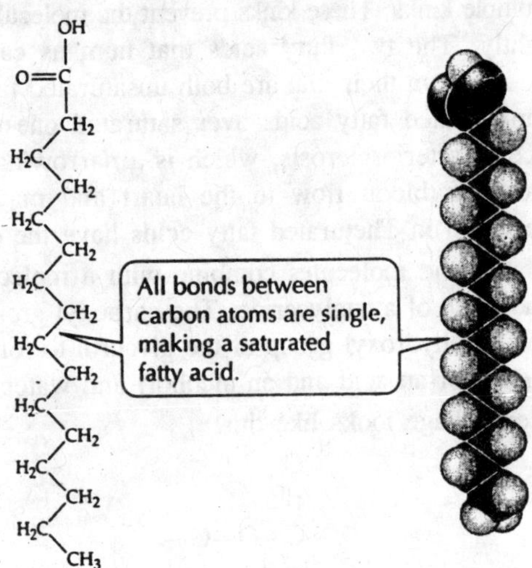

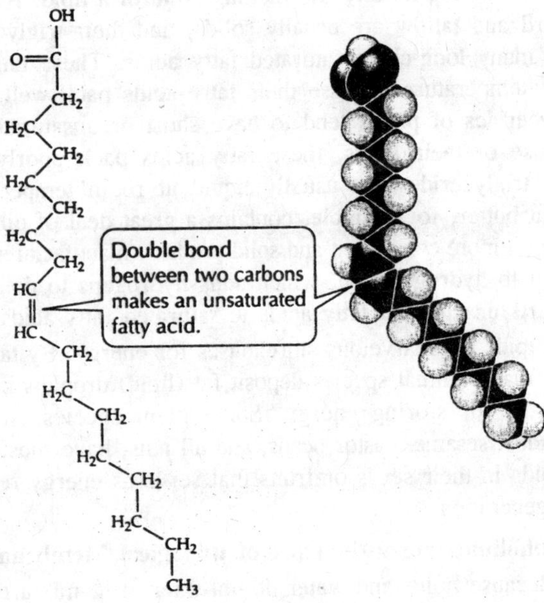

Fig. 3.2. Fatty acids.

linoleic acid, that have more than one double bond are polyunsaturated and have multiple kinks. These kinks prevent the molecules from packing together tightly. The two fatty acids that humans cannot synthesize and must obtain from their diet are both unsaturated fatty acids. Diets favouring unsaturated fatty acids over saturated ones tend to reduce the incidence of arteriosclerosis, which is a narrowing of the arterial wall that restricts blood flow to the heart and may lead to heart attacks. Diets rich in Theturated fatty acids have the opposite effect.

Three fatty acid molecules combine with a molecule of glycerol to form a molecule of a triglyceride. The carboxyl group of each fatty acid reacts with a hydroxyl group of the glycerol to form an *ester* (the reaction product of an acid and an alcohol) and water.

The ester linkage looks like this:

$$\begin{array}{c} O \\ \| \\ -C-O-C- \end{array}$$

The three fatty acids in one triglyceride molecule need not all have the same length, nor do they all have to be either saturated or unsaturated. The kinks associated with double bonds are important in determining the fluidity and melting point of a lipid. Animal fats such as lard and tallow are usually solids, and their triglycerides tend to have many long-chain saturated fatty acids. These fats are solids at room temperature because their fatty acids pack well together. The triglycerides of plants tend to have short or unsaturated fatty acids. Because of their kinks, these fatty acids pack poorly together and these triglycerides are usually liquid at room temperature. Natural peanut butter, for example, contains a great deal of oil. However, to market a more convenient and solid product, manufacturers often subject the oil to hydrogenation, which adds hydrogens to double bonds and converts unsaturated fatty acids to saturated fatty acids.

Lipids are marvelous storehouses for energy. By taking in excess food, many animal species deposit fat (lipid) droplets in their cells as a means for storing energy. Some plant species, such as olives, avocados, sesame, castor beans, and all nuts, have substantial amounts of lipids in their seeds or fruits that serve as energy reserves for the next generation.

Phospholipids Form the Core of Biological Membranes

Because lipids and water do not interact, a mixture of water and lipids forms two distinct phases. Many biologically important substances—such as ions, sugars, and free amino acids—that are soluble in water are insoluble in lipids. These two properties—water solubility and lipid insolubility—have proved essential to distinguishing the interior of cells from their external environment.

Suppose that you must design water-filled compartments, separated from each other and from their environment by barriers that limit the passage of materials. Given the properties of lipids, a seemingly effective way to accomplish this task is to use membranes that contain a special class of lipid called the phospholipid. This is the system that has evolved in nature. Molecular traffic within an organism or into and out of its compartments is constrained by the properties of the lipid portion of the surrounding membrane. Compounds that dissolve readily in lipids can move rapidly through biological membranes, but compounds that are insoluble in lipids are prevented from passing through the membrane or must be transported across the membrane by specific proteins.

Like triglycerides, phospholipids have fatty acids bound to glycerol by ester linkages. In phospholipids, however, any one of several phosphate-containing compounds may replace one of the fatty acids. Many phospholipids are important constituents of biological membranes. If you think carefully about the structure and properties of phospholipids, you will find it easy to understand how they are oriented in membranes. The phosphate functional group has negative electric charges, so this portion is hydrophilic, attracting polar water molecules. But the two fatty acids are hydrophobic, so they are pushed together by water.

In a biological membrane, phospholipids line up in such a way that the nonpolar, hydrophobic "tails" pack tightly together to form the interior of the membrane, and the phosphate-containing "heads" face outward (some to one side of the membrane and some to the other), where they interact with water, which is excluded from the interior of the membrane. The phospholipids thus form a bilayer, a sheet two molecules thick. Because membranes are so important.

Because the word "lipid" defines compounds in terms of their solubility rather than their structural similarity, a great variety of different chemical structures are included as lipids. The next two lipid classes we'll discuss—the carotenoids and the steroids—have chemical structures very different from the structures of triglycerides and phospholipids and from the structures of each other.

Carotenoids Trap Light Energy

The *carotenoids* are a family of light-absorbing pigments found in plants and animals. Betacarotene (β-carotene) is one of the pigments that traps light energy in leaves during photosynthesis. It is the β-carotene in plants that senses light and causes their parts to grow toward or away from the light. In humans, a molecule of β-carotene can be broken down into two vitamin A molecules, from which we make the pigment rhodopsin, which is required for vision. Carotenoids are responsible for the colour of carrots, tomatoes, pumpkins, egg yolks, and butter.

Steroids are Signal Molecules

The *steroids* are a family of organic compounds whose multiple rings share carbons. Some steroids are important constituents of membranes. Others are hormones, chemical signals that carry messages from one part of the body to another. *Testosterone* and the *estrogens* are steroid hormones that regulate sexual development in vertebrates. *Cortisol* and related hormones play many regulatory roles in the digestion of carbohydrates and proteins, in the maintenance of salt balance and water balance, and in sexual development.

Cholesterol is synthesized in the liver and contributes to the structure of some cellular membranes. It is the starting material for making testosterone and other steroid hormones, as well as the bile salts that help break down dietary fats so that they can be digested.

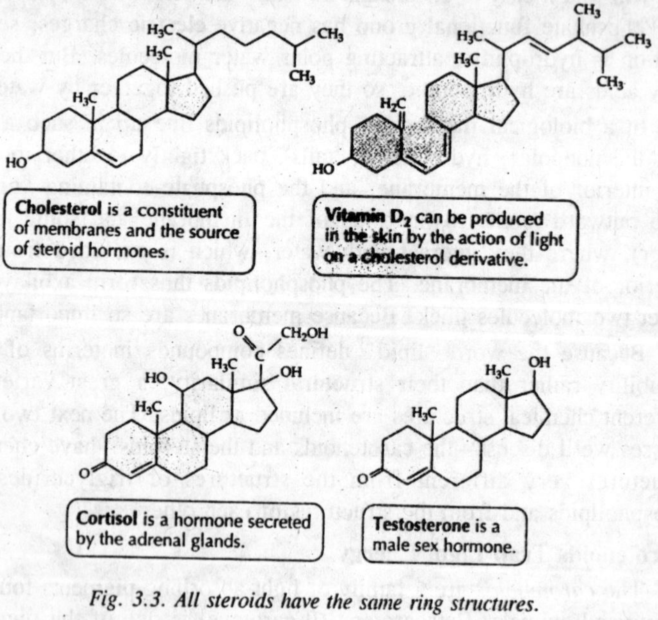

Cholesterol is a constituent of membranes and the source of steroid hormones.

Vitamin D_2 can be produced in the skin by the action of light on a cholesterol derivative.

Cortisol is a hormone secreted by the adrenal glands.

Testosterone is a male sex hormone.

Fig. 3.3. All steroids have the same ring structures.

We absorb cholesterol from foods such as milk, butter, and animal fats. When we have too much cholesterol in our blood, it is deposited in our arteries (along with other substances), a condition that may lead to arteriosclerosis and heart attack.

Some Lipids are Vitamins

A large group of fat-soluble substances, including the carotenoids and steroids, are synthesized by covalent linking and chemical modification of isoprene to form a series of isoprene units:

$$CH_2 = \overset{\overset{\displaystyle CH_3}{|}}{C}\text{—}CH = CH_2$$

The fat-soluble vitamins (A, D, E, and K) are formed in this manner by plants and bacteria. These substances are not synthesized by humans and must be acquired from dietary sources. *Vitamin A* is formed from β-carotene found in green and yellow vegetables. Among other roles, vitamin A is directly involved in the reception of light by our eyes. Deficiency in vitamin A leads to dry skin, eyes, and internal body surfaces; retarded growth and development; and night blindness, which is a diagnostic symptom for the deficiency. *Vitamin D* regulates the absorption of calcium from the intestines. It is necessary for the proper deposition of calcium in bones; a deficiency of vitamin D can lead to rickets, a bonesoftening disease.

Vitamin E is not a single vitamin, but a group of related lipids that seem to protect cells from damaging effects of oxidation-reduction reactions. These lipids appear to have an important role in preventing unhealthy changes in the double bonds in the unsaturated fatty acids of membrane phospholipids. Commercially, vitamin E is added to some foods to slow spoilage. *Vitamin K* is found in green leafy plants and is also synthesized by bacteria normally present in the human intestine. This vitamin is essential to the formation of blood clots. Predictably, a deficiency of vitamin K leads to slower clot formation and potentially fatal bleeding from a wound.

Because of their insolubility in water, in the body some lipids may require water-soluble carrier proteins in order to be transported in the blood, which is mostly water. Among the lipids, only triglycerides and phospholipids assemble into large structures in cells: triglycerides form droplets of stored fat, and phospholpids form the bilayers of membranes. But in spite of their size, these droplets and bilayers are not usually considered macromolecules.

Now that we have seen the macromolecule-like structures that lipids form, let's turn to the large structures that constitute true macromolecules.

Macromolecules: Giant Polymers

As noted earlier, *macromolecules* are giant *polymers* constructed by the covalent linking of smaller molecules called *monomers*. These monomers may or may not be identical, but they always have similar chemical structures. Molecules with molecular weights exceeding 1,000 are usually considered macromolecules, and the polysaccharides, proteins, and nucleic acids of living systems certainly fall into this category.

In the cell, each type of macromolecule performs some combination of a diversity of functions: energy storage, structural support, protection, catalysis, transport, defense, regulation, movement, and heredity. These roles are not necessarily exclusive. For example, both carbohydrates and proteins can play structural roles-supporting and protecting cells and organisms. However, only nucleic acids specialize in information and function as hereditary material, carrying both species and individual traits from generation to generation.

The functions of macromolecules are directly related to their shapes and the chemical properties of their monomers. Some macromolecules, such as catalytic and defensive proteins, fold into compact spherical forms with surface features that make them watersoluble and capable of intimate interaction with other molecules. Other proteins and carbohydrates form long, fibrous systems that provide strength and rigidity to cells and organisms. Still other long, thin assemblies of proteins can contract and cause movement.

Because macromolecules are so large, they contain many different functional groups. For example, a large protein may contain hydrophobic, polar, and charged groups. These groups give specific properties to local sites on a macromolecule. As we will see, this diversity of properties determines the shapes of macromolecules and their interactions with both other macromolecules and smaller molecules.

Macromolecules Form by Condensation Reactions

The polymers of living things are constructed by a series of reactions called *condensation reactions*, or dehydration hydration reactions (both words refer to the loss of water). Condensation reactions covalently bond monomers. Water is lost in the reaction:

$$A\text{—}H + B\text{—}OH \rightarrow A\text{—}B + H_2O$$

A—H is a molecule with a reactive hydrogen; B—OH is a molecule with a hydroxyl group. The products of the reaction are A—B and H_2O. The atoms that make up the water molecule are derived from the reactants: one hydrogen atom from one reactant, and an oxygen atom and the other hydrogen atom from the other reactant. Condensation reactions are not limited to polymer formation; for example, we encountered them in the linkage (esterification) of fatty acids to glycerol.

The condensation reactions that produce the different kinds of macromolecules differ in detail, but in all cases polymers will form only if energy is added to the system. In living systems, specific energy-rich molecules supply the energy, and there are additional steps to the reaction. The reverse of a condensation reaction is a *hydrolysis reaction* (*hydro-*, "water"; *-lysis,* "breakage"). These reactions digest polymers and produce monomers. Water reacts with the bonds that link the monomers together, and the products are free monomers. The elements of the reactant H_2O become part of the products. Hydrolysis reactions of all sorts are very important in cellular functioning and are not limited to the digestion of polymers.

Carbohydrates: Sugars and Sugar Polymers

Carbohydrates are a diverse group of compounds based on the general formula CH_2O. Some are relatively small, with molecular weights less than 100. Others are true macromolecules, with molecular weights of hundreds of thousands. There are four categories of biologically important carbohydrates. *Monosaccharides* (*mono-*, "one"; *saccharide,* "sugar") such as glucose or fructose—are the monomers out of which the larger forms are constructed. *Disaccharides* (*di-*, "two") consist of two monosaccharides. *Oligosaccharides* (*oligo-*, "several") have several monosaccharides (3 to 20). *Polysaccharides* (*poly-*, "many")-such as starch, glycogen, and cellulose—are composed of hundreds of thousands of glucose units.

The relative proportions of carbon, hydrogen, and oxygen indicated by the general formula for carbohydrates, CH_2O, is true for monosaccharides. However, for disaccharides, oligosaccharides, and polysaccharides, these proportions differ slightly from this general formula because two hydrogens and an oxygen are lost during the condensation reactions.

Monosaccharides are Simple Sugars

All living cells contain the monosaccharide glucose, whose formula is $C_6H_{12}O_6$. Green plants produce glucose by photosynthesis, and other

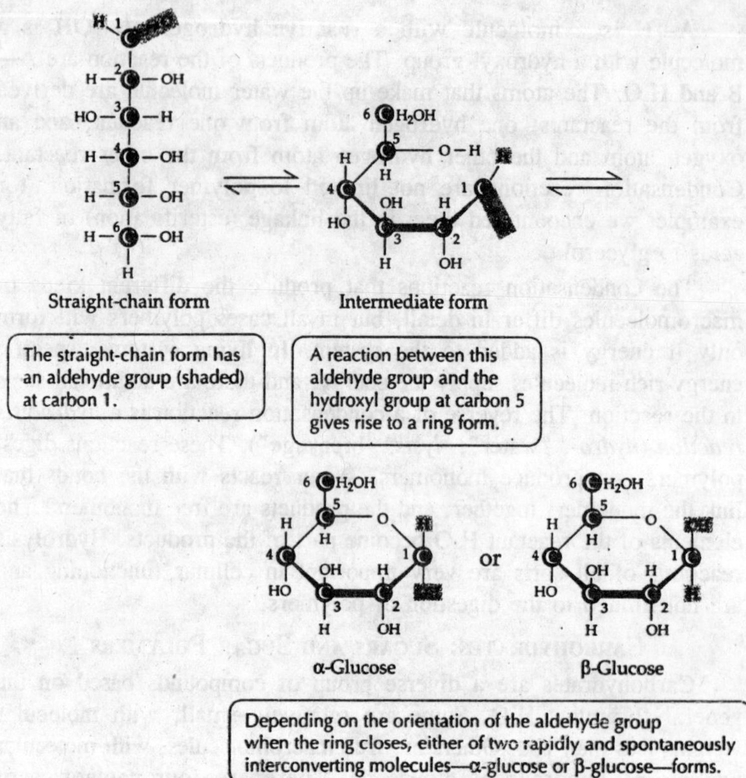

Fig. 3.4. Glucose: from one form to the other.

organisms acquire it directly or indirectly from plants. Cells use glucose as an energy source, changing it through a series of reactions that release stored energy and produce water and carbon dioxide.

Two forms of glucose, the straight chain and the ring, exist in equilibrium with each other when dissolved in water, but the ring form predominates (>99%). The two distinct ring forms (α and β-glucose) differ only in the placement of the —H and —OH attached to carbon 1. Most of the monosaccharides found in living systems belong to the D series of optical isomers. But there are structural isomers-composed of the same kinds and numbers of atoms, but with the atoms combined differently in each. All *hexoses* (*hex-*, "six"), a group of structural isomers, have the formula $C_6H_{12}0_6$. Included among the hexoses are fructose (so named because it was first found in fruits), mannose, and galactose.

Three-carbon sugar

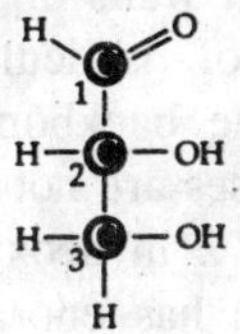

Glyceraldehyde is the smallest sugar and exists only as the straight-chain form.

Five-carbon sugars

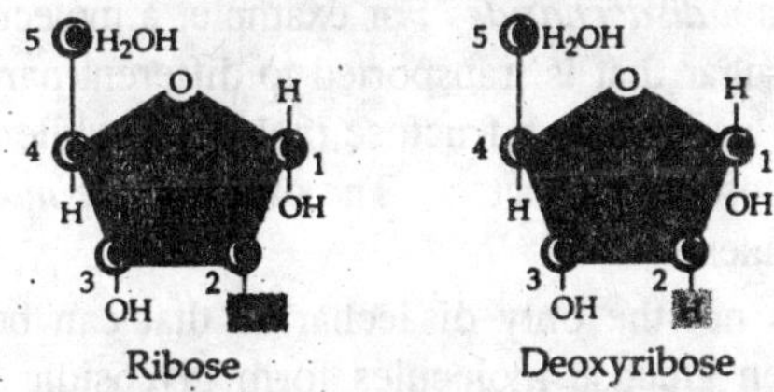

Ribose and deoxyribose each have five carbons, but very different chemical properties and biological roles.

Six-carbon sugars

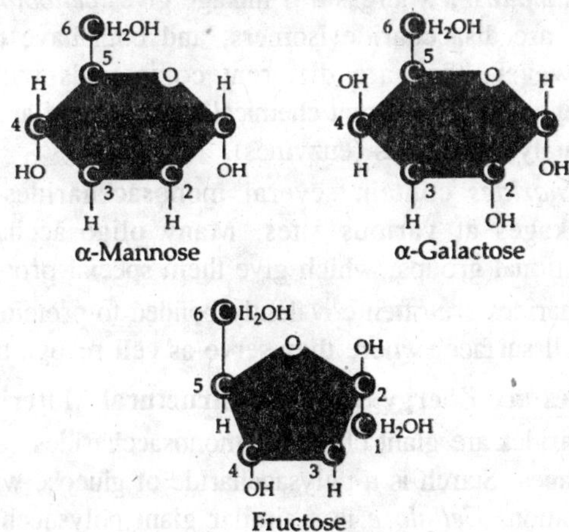

These hexoses are isomers. All have the formula $C_6H_{12}O_6$–but each has distinct chemical properties and biological roles.

Fig. 3.5. Monosaccharides are simple sugars.

Pentoses (*pent-*, "five") are five-carbon sugars. Some pentoses are found primarily in the cell walls of plants, as are several of the hexoses. Two pentoses are of particular importance: *Ribose* and *deoxyribose* form part of the backbones of RNA and of DNA, respectively. These two pentoses are not isomers; rather, one oxygen atom is missing from carbon 2 in deoxyribose (de, "absent"). The absence of this oxygen atom has enormous consequences for the functional distinction of RNA and DNA.

Glycosidic Linkages bond Monosaccharides Together

Monosaccharides are covalently bonded together by condensation reactions that form *glycosidic linkages*. Such a linkage between two simple sugars forms a *disaccharide*. For example, a molecule of *sucrose* (table sugar)—the sugar that is transported to different parts of plants—is formed from a glucose and a fructose molecule, while *lactose* (milk sugar) contains glucose and galactose. The disaccharide *maltose* contains two glucose molecules.

But maltose is not the only disaccharide that can be made from two glucoses. When glucose molecules form glycosidic linkages, the disaccharide product must be one of two types: α-linked or β-linked, depending on whether the molecule that bonds by its 1 carbon is α-glucose or β-glucose. An α linkage with carbon 4 of a second glucose molecule gives *maltose*, whereas a β linkage gives *cellobiose*. Maltose and cellobiose are disaccharide isomers, and both have the formula $C_{12}H_{22}O_{11}$. However, they are different compounds with different properties. They undergo different chemical reactions and are recognized by different catalytic proteins (enzymes).

Oligosaccharides contain several monosaccharides linked by glycosidic linkages at various sites. Many oligosaccharides gain additional functional groups, which give them special properties.

Oligosaccharides are often covalently bonded to proteins and lipids on the outer cell surface, where they serve as cell recognition signals.

Polysaccharides are Energy Stores or Structural Materials

Polysaccharides are giant chains of monosaccharides connected by glycosidic linkages. Starch is a polysaccharide of glucose with linkages in the a orientation. *Cellulose* is a similar giant polysaccharide made up solely of glucose, but its individual units are connected by β linkages instead of a linkages. Cellulose is the predominant component of plant cell walls and is by far the most abundant organic compound on this planet. Both starch and cellulose are composed of nothing but glucose, but their biological functions and chemical and physical properties are entirely different.

Starch can be more or less easily degraded by the actions of chemicals or catalytic proteins (enzymes). Cellulose, however, is very stable because of its β glycosidic linkages. Thus starch is a good storage form that can be easily degraded to supply glucose for energy producting reactions, while cellulose is an excellent structural component that can withstand harsh environmental conditions without changing.

Starch is not just one chemical substance, but large family of giant molecules of broa structure. All starches are large polymers of glucose with a linkages, but different starches can be distinguished by the amount of branching between carbons 1 and 6. Some starches are highly branched; others are not. Plant starches, called *amylose*, are not highly branched. The polysaccharide glycogen, which stores glucose in animal livers and muscles, is highly branched.

What do we mean when we say that starch and glycogen are storage compounds for energy? Very simply, these compounds can readily be hydrolyzed to yield glucose monomers. Glucose, in turn, can enter into a series of reactions that liberate its stored energy in forms that can be used for cellular activities. However, glucose can also serve the cell's need for raw materials—that is, carbon atoms. Glucose can undergo other chemical reactions that disassemble parts of the molecule and rearrange its carbon atoms to form the skeletons of other compounds needed by cells. Glycogen and starch are thus storage depots for carbon atoms as well as for energy.

Derivative Carbohydrates Contain other Elements

Sometimes carbohydrates are modified by chemical changes in their structure or by the addition of functional groups such as phosphate and amino groups, thus becoming *derivative carbohydrates*. For example, carbon 6 in glucose may be oxidized from $-CH_2OH$ to a carboxyl group ($-COOH$), producing glucuronic acid. Or a phosphate group may be added to one or more of the $-OH$ sites. Some of these *sugar phosphates*, such as fructose 1,6-bisphosphate, are important intermediates in cellular energy reactions. When an amino group is substituted for an $-OH$ group, *amino sugars* such as glucosamine and galactosamine are produced. Galactosamine is a major component of cartilage, the material that forms caps on the ends of bones and stiffens the protruding parts of the ears and nose. A derivative of glucosamine produces the polymer *chitin*, which is the principal structural polysaccharide in the skeletons of insects, crabs, and lobsters, as well as in the cell walls of fungi. Fungi and insects (and their relatives) constitute more than 80 percent of the species ever described, and chitin is one of the most abundant substances on Earth.

Proteins: Amazing Polymers of Amino Acids

Proteins are an extraordinary group of macromolecules. Constructed of amino acids, proteins are fascinating because they have such intricate and diverse structures and because they perform so many different functions for cells. Among the cellular functions of macromolecules listed earlier, only energy storage and heredity are not functions of proteins. Proteins are involved in structural support, protection, catalysis, transport, defense, regulation, and movement. Of particular importance are the catalytic proteins, called *enzymes*, that increase the rates of chemical reactions in cells. In general, each chemical reaction requires a different enzyme, because proteins show great specificity for the smaller molecules with which they will interact.

Proteins range in size from the small RNA-digesting enzyme *ribonuclease A*, which has a molecular weight of 5,733 and 51 amino acid residues, to gigantic molecules such as the cholesterol transport protein *apolipoprotein B*, which has a molecular weight of 513,000 and 4,636 amino acid residues. (The word "residue" is used for a monomer when it is part of a polymer.) Each of these proteins consists of one chain of amino acids folded into a specific three-dimensional shape that is required for protein function. Some proteins have more than one polymer chain. For example, the oxygen-carrying protein *hemoglobin* has four chains that are folded separately and associate together to make the functional protein. The largest protein complex known has more than 40 separate chains of amino acids.

All of these different proteins have a characteristic amino acid composition, but every protein contains neither all 20 amino acids nor an equal number of different amino acids. Nor is there a simple regular sequence in which the amino acids are linked. The diversity in amino acid content and sequence is the source of the diversity in protein structures and functions. In some proteins, different kinds of chemical substances, called *prosthetic groups*, may be attached to the protein. These prosthetic groups include carbohydrates, lipids, phosphate groups, the iron-containing heme group, and metal ions such as copper and zinc. Whether they are small or large, and whether or not they have prosthetic groups, there is nothing "casual" about the structures of proteins. Proteins have specific three-dimensional shapes that are necessary for their specific functions.

To understand this stunning variety of functions, we must explore protein structure. First, we will examine the properties of the 20 amino acids and the characteristics of how they link to form proteins. Then we will systematically examine the four levels of protein structure and look at how a linear chain of amino acids is consistently folded into a compact three-dimensional shape.

Proteins are Composed of Amino Acids

The 20 different amino acids commonly found in proteins show a wide variety of properties. We considered the structure of amino acids and identified four different groups attached to a central carbon atom: a hydrogen atom, an amino group, a carboxyl group, and a side chain, or R group. Since amino acids are linked together by reactions between their amino and carboxyl groups, these groups are not exposed and do not give the protein its distinguishing properties and functional specificity-the side chains fill this role.

The *side chains* of amino acids are a protein's reactive groups and show a wide variety of chemical properties. Side chains control the function of a protein and contribute to its structure. The sequence of side chains determines how a protein folds into a three-dimensional shape (which we will discuss shortly). Although side chains are very important, they are often omitted from diagrams of proteins when the focus is on other aspects of structure. In these cases, the side chains are identified by the letter R (for "residue") and are sometimes called R groups.

One useful classification of amino acids is based on whether their side chains are electrically charged (+1, -1), polar (δ+, δ-), or nonpolar and hydrophobic. The five amino acids that have electrically charged side chains attract water and oppositely charged ions of all sorts. The four amino acids that have polar side chains tend to form weak hydrogen bonds with water and with other polar or charged substances. Eight amino acids have side chains that are nonpolar hydrocarbons or very slightly modified hydrocarbons. In the watery environment of the cell, the hydrophobic side chains may cluster together.

Three amino acids—cysteine, glycine, and proline—are special cases, although their side chains are generally hydrophobic. Two *cysteine* side chains, which have terminal —SH groups, can react to form a covalent bond in a *disulfide bridge* (—S—S—). Hydrogen bonds and disulfide bridges help determine how a protein chain folds. When cysteine is not part of a disulfide bridge, its side chain is very hydrophobic. The *glycine* side chain consists of a single hydrogen atom; thus glycines may fit into tight corners in the interior of a protein molecule, where a larger side chain could not fit. *Proline* differs from other amino acids because it possesses a modified amino group.

Peptide Linkages Covalently Bond Amino Acids Together

When amino acids polymerize, the carboxyl group of one amino acid reacts with the amino group of another, undergoing a condensation reaction that forms a peptide linkage. Gives a simplified description of the reaction. (In living cells, other molecules must activate the reactants in order for this reaction to proceed, and there are intermediate steps.) A linear polymer of amino acids connected by peptide linkages is a *polypeptide*. A *protein* is made up of one or more polypeptides.

At one end of the polypeptide is a free amino group. This end is the N terminus, named for the nitrogen atom in the amino group. At the other end of the polypeptide—the C terminus—is a free carboxyl group. The other amino and carboxyl groups are bound in peptide linkages. Thus a protein has direction. For example, the dipeptide glycine-alanine, in which glycine has the free amino group, differs from alanine-glycine, in which alanine has the free amino group. In cells, the synthesis of polypeptide chains begins with the N terminus.

In the peptide linkage, the C=O oxygen carries a slight negative charge (δ-), whereas the N—H hydrogen is slightly positive (δ+). This asymmetry of charge favours hydrogen bonding within the protein molecule itself and with other molecules, contributing to both the structure and the function of many proteins.

Primary Structure of a Protein is its Amino Acid Sequence

Protein structure is elegant and complex—so complex that it is described as consisting of four different levels: *primary*, *secondary*, *tertiary*, and *quaternary*. However, proteins are always linear chains; there are no branches. The precise sequence of amino acids in a polypeptide constitutes the primary *structure* of a protein. The *peptide backbone* of this primary structure consists of a repeating sequence of three atoms (—N—C—C—) from the amino group, the central carbon, and the carboxyl group of each amino acid.

In cells, the primary structure of a protein is dictated by the precise sequence of nucleotides in a linear segment of a DNA molecule. The elucidation of this relationship between DNA primary structure and protein primary structure was one of the triumphs of molecular biology.

The theoretical number of different proteins is enormous. Since there are 20 different amino acids, there are $20 \times 20 = 400$ distinct dipeptides, and $20 \times 20 \times 20 = 8{,}000$ different tripeptides. Imagine this process of multiplying by 20 extended to a protein made up of 100 amino acids (which is considered a small protein): There could be 20^{100} of these small proteins, each with its,own distinctive primary structure. How large is the number 20^{100}? In the entire universe there aren't that many electrons!

At the higher levels of protein structure, local coiling and folding give the final functional shape of the molecule, but all of these levels derive from the primary structure. The different properties associated with a precise sequence of amino acids determine how the protein can twist and fold. By twisting and folding, each protein adopts a specific stable structure that distinguishes it from every other protein.

Secondary Structure of a Protein Requires Hydrogen Bonding

Although the primary structure of each protein is unique, the secondary structure of many different proteins may be the same. A protein's *secondary structure* consists of regular, repeated patterns in different regions of a polypeptide chain. One type of secondary structure, the *α helix* (alpha helix), is a right-handed coil that is "threaded" in the same direction as a standard wood screw. The amino acid side chains extend outward from the peptide backbone of the helix.

This helical structure of a polypeptide chain results from hydrogen bonds between elements of the peptide bonds that are distributed along the back bone of the chain. Hydrogen bonds form between the slightly positive hydrogen of the N—H of one peptide bond and the slightly negative oxygen of the C=O of another peptide bond. When this pattern of hydrogen bonding is established repeatedly over a segment of the protein, it stabilizes the twisted form, resulting in an a helix. However, the ability of a protein to form an a helix depends on its primary structure. Amino acids with larger side chains that distort the coil or otherwise prevent the formation of the necessary hydrogen bonds will keep the a helix from forming.

Alpha helical secondary structure is particularly evident in the fibrous structural proteins called *keratins*. Keratins constitute many of the protective materials found in mammals and birds, such as fingernails and claws, skin, hair, wool, and feathers. Hair can be stretched because this stretching requires that only hydrogen bonds in an a helix, and not covalent bonds, be broken; when the tension on the hair is released, both the helix and the hydrogen bonds reform.

Another type of secondary structure, the β pleated sheet, is found in the protein silk. In silk, rather than being coiled, the protein chains are almost completely extended and lie next to one another, stabilized by hydrogen bonds between the elements of the peptide linkages. The R pleated sheet may be found between separate polypeptide chains, as in silk, or between different regions of the same polypeptide that is bent back on itself. Many enzymes contain regions of a helix and of β pleated sheet in the same polypeptide chain.

Tertiary Structure of a Protein is Formed by Bending and Folding

In most proteins, to establish the compact structure the polypeptide chain must be bent at specific sites and folded back and forth. The resulting overall shape of a protein is its tertiary structure. Although the a helices and β pleated sheets contribute to the tertiary structure, more frequently only limited portions of the molecule have these secondary structures, and large regions consist of structures unique to a particular protein.

A complete description of the tertiary structure specifies the location of every atom in the molecule in three-dimensional space, in relation to all the other atoms. Bear in mind that this tertiary structure and the secondary structure derive from the protein's primary structure. If lysozyme is heated carefully, causing only the tertiary structure to break down, the protein will return to its normal tertiary structure when it cools. The only information needed to specify the unique shape of the lysozyme molecule is the information contained in its primary structure.

Hemoglobin is the protein that transports oxygen (O_2) from the lungs to the tissues. In muscle tissue, hemoglobin delivers O_2 to myoglobin, a smaller but similar protein, for storage. The hemoglobin molecule consists of four similar polypeptide chains that have tertiary structures made up almost entirely of a helices but that lack any disulfide bridges. The helical segments bend and fold against each other, forming a pocket that encloses a *heme group*, an ironcontaining prosthetic group that binds O_2. The stabilizing interaction of hydrophobic side chains on the inner sides of the a helices helps ensure that the helices fold against one another correctly as each polypeptide assumes its tertiary structure.

Quaternary Structure of a Protein Consists of Subunits

As we mentioned earlier, some proteins have two or more polypeptide chains folded into their own unique tertiary structures.

Each of these folded polypeptides is considered a protein subunit. *Quaternary structure* results from the ways in which these protein subunits fit together and interact; it can be illustrated by hemoglobin. Hydrophobic interactions, hydrogen bonds, and ionic bonds all help hold the four subunits together to form the functional hemoglobin molecule. As the hemoglobin molecule takes up one O_2 molecule, the four subunits shift their relative positions slightly, changing the quaternary structure. Ionic bonds are broken, exposing buried side chains that enhance the binding of additional O_2 molecules.

Each subunit of hemoglobin is folded like a myoglobin molecule, suggesting that both hemoglobin and myoglobin are evolutionary descendants of the same oxygen-binding ancestral protein. But on the surfaces where its subunits come in contact with each otherregions that on myoglobin are exposed to aqueous surroundings and are hydrophilic-hemoglobin has hydrophobic side chains. Again, the chemical nature of side chains on individual amino acids determines how the molecule folds and packs in three dimensions.

Molecular Chaperones Help Shape Proteins

The primary structure of a protein constrains the secondary, tertiary, and quaternary structures (if subunits exist). By determining the primary structure, DNA also determines the higher levels of structure. However, other factors also affect the tertiary structure that is required for proper protein function.

Elevated temperatures, pH changes, or altered salt concentrations can cause a protein to adopt a different, biologically inactive tertiary structure. Biological function depends on a specific three-dimensional structure. Increased temperature causes more rapid molecular movement and thus can break weak hydrogen bonds and hydrophobic interactions. Altered pH can change the pattern of ionization of carboxyl and amino groups in the side chains of amino acids, thus disrupting the pattern of ionic attractions and repulsions that contribute to normal tertiary structure.

The loss of appropriate tertiary structure is called *denaturation*, and it is always accompanied by a loss of the normal biological function of the protein. Denaturation can be caused by heat or high concentrations of polar substances such as urea that disrupt the hydrogen bonding that is crucial to protein structure. Nonpolar solvents may also disrupt normal structure. Usually denaturation is irreversible, particularly if many denatured proteins interact in random nonspecific ways. However, in some cases denaturation is reversible, and upon return to normal environmental conditions, the protein may return to its active form, as does lysozyme. This return is called *renaturation*.

How can such a complicated molecule spontaneously fold into its normal tertiary structure? Biologists are just beginning to be able to predict how a protein will fold, given its primary structure. The study of protein folding has a long way to go. The situation in the living cell is complex-even more complex than we first imagined. The protein doesn't form all at once. In a human cell it may take minutes from the beginning of synthesis until the entire protein squeezes into the compartment where it folds-and things are very crowded in that compartment. There, other proteins present inappropriate potential partners for interaction (in the form of amino acid side chains).

A special group of proteins, called the *chaperone proteins*, at least in part help prevent newly formed proteins from reacting inappropriately. The chaperones attach to a new protein while it is forming and protect it from interactions with other proteins until the new molecule has its correct shape. Most chaperone proteins can act on many different forming proteins. Some act on only a few, and others may act on only a single, specific protein.

What if the chaperone proteins fail to do their job, or if for other reasons a protein folds abnormally? Evidence suggests that abnormal protein folding underlies certain infectious diseases, including mad cow disease and, in humans, Creutzfeldt-Jakob disease. Alzheimer's disease may also result from protein misfolding.

Nucleic Acids: Informational Macromolecules

The nucleic acids are linear polymers specialized for the storage, transmission, and use of information. There are two types of nucleic acids: DNA (deoxyribonucleic acid) and RNA (ribonucleic acid). DNA molecules are giant polymers that encode hereditary information and pass it from generation to generation (through reproduction). The information encoded in DNA is also used to make specific proteins through the intermediate RNA. RNA molecules of various types copy the information in segments of DNA to specify the sequence of amino acids in proteins. Information flows from DNA to DNA in reproduction. But for nonreproductive activities of the cell, information flows from DNA to RNA to proteins, which ultimately carry out these functions. What compositions, structures, and properties of nucleic acids permit them to play these fundamental roles in living systems?

Nucleic Acids have Characteristic Structures and Properties

Nucleic acids are composed of monomers called *nucleotides*, each of which consists of a pentose sugar, a phosphate group, and a nitrogen-containing base-either a pyrimidine or a purine. Molecules consisting of a pentose sugar and a nitrogenous base, but no phosphate group, are called nucleosides. In DNA, the pentose sugar is deoxyribose, which differs from the ribose found in RNA by only one oxygen atom.

In both RNA and DNA, the backbone of the molecule consists of alternating sugars and phosphates (—sugar—phosphate—sugar—phosphate—). Bases are attached to the sugars and project from the chain. The nucleotides are joined by covalent bonds in what are called *phosphodiester* linkages between the sugar of one nucleotide and the phosphate of the next ("-diester" refers to the two bonds formed by reacting —OH groups with acidic phosphate groups). The phosphate groups link carbon 3' in one ribose to carbon 5' in the adjacent ribose.

Most RNA molecules are *single-stranded*, consisting of only one polynucleotide chain. DNA, however, is usually *double-stranded*; it has two polynucleotide chains held together by hydrogen bonding between their complementary nitrogenous bases. The two strands of DNA run in opposite directions. You can see what this means by drawing an arrow through the phosphate group from carbon 3' to carbon 5' in the next ribose. If you do this for both strands, the arrows point in opposite directions. This antiparallel orientation is necessary for the strands to fit together.

Uniqueness of a Nucleic Acid Resides in its Base Sequence

Only four nitrogenous bases—and thus only four nucleotides—are found in DNA. The DNA bases and their abbreviations are: adenine (A), cytosine (C), guanine (G), and thymine (T). A key to understanding the structures and functions of nucleic acids is the principle of *complementary base pairing* through hydrogen bond formation. In double-stranded DNA, adenine and thymine always pair (AT), and cytosine and guanine always pair (CG).

Base pairing is complementary because of two factors: the corresponding sites for hydrogen bonding and the molecular sizes of the paired bases. Adenine and guanine are both purines consisting of two fused rings. Thymine and cytosine are both pyrimidines consisting of only one ring. The pairing of a large purine with a small pyrimidine ensures a stable and consistent dimension to the double-stranded molecule of DNA. We'll discuss in more detail how complementary strands of DNA separate and are faithfully copied.

Ribonucleic acids also have four different monomers, but the nucleotides differ from those of DNA. In RNA the nucleotides are termed ribonucleotides. They contain ribose rather than deoxyribose, and instead of the base thymine, RNA uses the base uracil. The other three bases are the same as in DNA.

Although RNA is generally single-stranded, complementary hydrogen bonding between ribonucleotides can take place. These bonds play important roles in determining the shapes of some RNA molecules and in the associations between RNA molecules during protein synthesis. During the DNA-directed synthesis of RNA, complementary base pairing also takes place between ribonucleotides and the bases of DNA. In RNA, guanine and cytosine pair as in DNA, but adenine pairs with uracil. Adenine in an RNA strand can pair either with uracil (in another RNA strand) or with thymine (in a DNA strand).

The three-dimensional appearance of DNA is strikingly regular. Through hydrogen bonding, the two complementary polynucleotide strands pair and twist to form a double helix. Compared to the complex and varied tertiary structures of different proteins, this formation seems amazingly regular. But this structural contrast makes sense in terms of the functions of these two classes of macromolecules.

DNA is a purely informational molecule. *The information in DNA is encoded in the sequence of bases carried in its chains.* This sequence is, in a sense, like the tape of a tape recorder. The message must be read easily and reliably, in a specific order. A uniform molecule like DNA can be interpreted by standard molecular machinery, and any cell's machinery can read any molecule of DNA-just as a tape player can play any tape of the right size.

Table 3.1. Distinguishing RNA from DNA

Nucleic acid	*Sugar*	*Bases*
RNA	Ribose	Adenine Cytosine Guanine Uracil
DNA	Deoxyribose	Adenine Cytosine Guanine Thymine

Proteins, on the other hand, have good reason to be so varied. In particular, different enzymes must recognize their own specific "target" molecules, They do this by having a unique three-dimensional form that can match at least a portion of the surface of their target molecules. In other words, structural diversity in the molecules with which enzymes react requires corresponding diversity in the structure of the enzymes themselves. *In DNA the information is in the sequence of bases; in proteins the information is in the shape.*

DNA is a Guide to Evolutionary Relationships

Because DNA carries hereditary information between generations, a series of DNA molecules with changes in base sequences stretches back- through time. Of course, we cannot study all of these DNA molecules, because many of their organisms have become extinct. However, we can study the DNA of living organisms, some of which are judged to belong to lineages that are more ancient than those of other living forms. Comparisons and contrasts of these DNAs add to the evolutionary record.

Closely related living species should have more similar base sequences than species judged by other criteria to be more distantly related. Indeed this is the case. The examination of base sequences confirms many of the evolutionary relationships that have been inferred from the study of microscopic or macroscopic structures or studies of biochemistry and physiology. For example, the closest living relative of humans (*Homo sapiens*) is the chimpanzee (genus *Pan*), which shares more than 98 percent of its DNA base sequence with human DNA.

Confirmation of well-established evolutionary relationships gives credibility to the use of DNA to elucidate relationships when studies of structure are not possible or are not conclusive. For example, DNA studies revealed a close evolutionary relationship between starlings and mocking-birds that was not expected on the basis of studies of anatomy or behaviour. DNA studies support the division of the prokaryotes into two kingdoms (Eubacteria and Archaebacteria). Each of these two groups of prokaryotes is as distinct from the other as either is from the other four kingdoms into which living things are classified. In addition, DNA comparisons support the hypothesis that certain subcellular compartments of eukaryotes (the organelles called *mitochondria* and *chloroplasts*) evolved from early bacteria that established a stable and mutually beneficial way of life inside larger cells.

Interactions of Macromolecules

We have been treating the classes of macromolecules as if each were completely separate from the others. In cells, however, certain macromolecules of different classes may be covalently bonded to one another. Proteins with attached oligosaccharides are called *glycoproteins* (*glyco-*, "sugar"). The specific oligosaccharide chain determines the placement of a glycoprotein within the cell.

Other carbohydrate chains bind to lipids, resulting in *glycolipids*, which reside in the cell surface membrane, with the carbohydrate chain extending out into the cell's environment. In humans, the substances that determine the ABO blood types are carbohydrates attached to either proteins or lipids. When a human body cell becomes cancerous, the glycolipids and glycoproteins on the cell surface are modified, and these changes serve as a recognition signal for the body defenses to destroy the abnormal cell.

Not all the associations among macromolecules or between a macromolecule and a smaller molecule are covalent. Weaker linkages such as ionic bonds, hydrogen bonds, and even van der Waals forces may also be involved. Enzymes bind to their reactants using a variety of weak bonds. Ionic bonds link proteins to DNA, forming *nucleoproteins* that regulate the activities of DNA. Still other proteins, in combination with cholesterol and other lipids, form *lipoproteins*. The lipoproteins make it possible to move very hydrophobic lipids through water-rich environments such as the blood and tissue fluid and deliver cholesterol to all the body's cells.

4

MACROMOLECULAR ENGINEERING

In introducing the concept of molecular manufacturing, examples of molecular devices and processes from biology were used to make the case for physical feasibility, and an analysis of the feasibility of protein engineering was used to make the case for technological feasibility. Although this use of examples from molecular biology has since been superseded by more direct design and analysis, these examples still illustrate how the engineering of biomolecules could enable the construction of a first generation of molecular machines.

MACROMOLECULAR OBJECTS

DNA, RNA, and Protein

The macromolecules of central concern in biotechnology and molecular biology are DNA, RNA, and protein. Genetic engineering now includes powerful and general techniques for constructing DNA molecules; via RNA, these can direct the molecular machinery of bacterial cells to make protein molecules.

DNA ordinarily functions as an information molecule, but N. Seeman has demonstrated that it can also be used to build structures with both branches and loops, including a molecule with the connectivity of a cube. The ability of complementary segments of DNA to bind to one another provides a powerful, readily controllable mechanism for guiding the Brownian assembly of molecules. Hybrid structures of DNA and other macromolecules may prove particularly attractive.

RNA in living cells has diverse functions; some are structural and chemical (e.g., in the ribosome) and some informational (e.g., in the messenger RNA that directs ribosomal synthesis of proteins). Like

DNA, it can fold and bind to match up complementary segments, forming a double helix.

Molecular machines in living cells are chiefly made of protein. Unlike pure DNA and RNA structures, protein structures can be dense, compact, and relatively stiff; further, their 20 amino acids give them greater chemical and structural versatility than is found in structures consisting only of the four nucleotides of DNA or of RNA. These characteristics suggest that proteins will playa central role if biotechnology is used to build nanomechanical systems.

Unlike DNA and RNA, however, protein chains do not bind to one another by pairwise matching of their monomers: no simple notion of complementarity can be used to predict or design the folding of a protein. Although nature demonstrates that proteins are adequate building blocks for systems of molecular machinery, and although biotechnology demonstrates that novel proteins can be constructed, the problem of protein design—of relating amino acid sequence to three-dimensional structure and function—has slowed progress toward protein-based macromolecular systems engineering.

Protein Folding: Prediction vs. Design

The "*protein folding problem*" has classically been defined as a *fold-prediction* problem, that is, predicting the three-dimensional folded conformation of a naturally occurring protein molecule given only its amino acid sequence (and no knowledge of the fold of a similar sequence). Despite continuing progress, no general solution to this problem has been found. It was long assumed that solving the fold-prediction problem was a prerequisite for solving the fold-design problem; this impression lingers, discouraging efforts in *de novo* protein engineering. Actually, these problems are quite distinct. In solving a design problem, one is free to bias all manipulable characteristics to favour a particular result, but in predicting a natural phenomenon, one faces whatever ambiguities nature happens to present. Based on this elementary observation, together with evolutionary considerations that predict marginal stability for natural proteins, split the "protein folding problem" into the distinct problems of fold design and fold prediction, arguing that fold design is easier. *De novo* design of globular proteins has since been pursued, yielding a recent landmark success which confirmed predictions that design could yield structures of unusual stability. Meanwhile, the ongoing failure to predict the folds of novel natural proteins confirmed the conclusion that fold design is both distinct from and easier than fold prediction.

Rational Design and Evolutionary Approaches

At present, the field of protein engineering divides into several main areas. Most protein engineering experiments modify one or more amino acid residues in a natural structure to alter protein stability or function. More ambitious experiments splice portions of known structures together to make new proteins. The most ambitious experiments, termed *de novo* protein engineering, make protein molecules that have no close resemblance to any found in nature. Experiments of the first sort have become routine; *de novo* protein engineering remains exceptional.

Recent years have seen the emergence of several techniques that develop new proteins not by design, but by an evolutionary process of variation, selection, and replication. The first group, monoclonal antibody technologies, exploits animal immune systems to develop protein molecules that bind a specific ligand molecule; by developing antibodies that bind a molecule resembling the transition state of a particular chemical reaction, it has proved possible to develop antibodies (*abzymes*) that catalyze the reaction. More recently, techniques for *in vitro* evolution of monoclonal antibody molecules manufactured by bacteria have been demonstrated. Systems that select among molecular objects by their affinity to a chromatographic column have been used to evolve specific binding affinities both in protein fragments exposed on viral surfaces and in isolated RNA fragments.

Evolutionary techniques can be used to develop protein molecules capable of binding (and altering) other molecules, so long as the desired property permits differential selection and does not interfere with the replication mechanism. If protein molecules are to be used as self-assembling building blocks in larger structures, developing proteins that bind to the correct surface sites is a basic requirement. Evolutionary techniques may provide means for developing such molecules, if methods can be found to select for strong binding at desired sites on a target structure. Methods for masking alternative sites, or for selecting *against* binding to alternative sites may prove useful in evolving the desired specificity.

Material and Device Properties

In designing machinery for molecular manufacturing, stiffness is a central concern. Different protein structures have widely differing moduli of elasticity (and are inhomogeneous and anisotropic as well); a Young's modulus of ~ 10 GPa nonetheless "seems to be common to various kinds of polymers of globular protein molecules". This falls within the range of moduli of polymeric materials such as polyethylene, 1 GPa; polystyrene, ~3 GPa: polyphenylene sulfide, 8 GPa; poly (chlorotrifluoroethylene), 14 GPa; and most kinds of wood, 8 to 12 GPa. Although a modulus of 10 GPa is adequate for many purposes, it is only 1% of the modulus of diamond: there is thus ample motivation to move to better materials.

Existing protein devices demonstrate forces, speeds, and frequencies that can be achieved by machinery constructed from biomolecules working in solution. The enzyme having the highest known operating frequency, catalase. decomposes hydrogen peroxide molecules at $4 \times 10^7\ s^{-1}$; bacterial DNA replication machinery can add nucleotides at $750\ s^{-1}$, and bacterial ribosomes can add amino acids at about $40\ s^{-1}$. The bacterial flagellar motor can rotate at almost 300 Hz, and can develop a torque of $\sim 2.5 \times 10^{-18}$ N·m (= 2.5 nN·nm). Several molecular linear motor (based on myosin, kinesin, and dynein) exert forces that have been estimated a 0.0005 to 0.0026 nN.

Development paths starting with biomolecular systems can exploit existing molecular machines and components. This advantage is reduced by the complexity, fragility, and specialization of existing biomolecular systems, and by the limited knowledge of many structures (for example, no motor mechanism has yet been characterized in atomic detail). Structural knowledge will accumulate (aided by new instruments), and protein engineering techniques might be used to adapt these mechanisms to other applications and make them more robust. Developments along several other pathways, however, will be compatible with hybrid strategies in which biomolecular and wholly synthetic components are combined. Accordingly, it seems unlikely that a *purely* biomolecular approach will be carried through.

Macromolecular Objects via Solution Synthesis

Motivation

Design becomes easier in systems that exhibit good functional modularity, and becomes more reliable when ample safety margins are possible; greater freedom of design can further both of these objectives. Alternatives to standard, genetically encoded proteins can provide a more tractable medium for the design of folded molecules with proteinlike functions, chiefly by providing better modularity and safety margins. Greater design freedom will permit novel strategies for stabilizing folds; estimates based on observed correlations between structure and free energy suggest that order-of-magnitude increases in the stability of the folded state can be achieved. Increased stability and rigidity can be expected to mitigate errors in molecular modeling and design, increasing the odds of success in a particular design. The family of approaches outlined in this section adheres closely to the protein model; attractive approaches can likely be found that are quite dissimilar from any biological model.

Constructing molecular objects poses challenges both in design and in synthesis, but one challenge can be traded off against the other. Alternatives to standard proteins typically sacrifice the convenience of biosynthesis in exchange for greater stability and reduced design difficulty. Protein design remains challenging chiefly because the structure of a globular protein emerges from a cooperative folding process dependent on a balance of weak, noncovalent forces in a large, flexible structure. At the other extreme of molecular size and flexibility, proteinlike receptors have been constructed from relatively small, rigid, polycyclic molecules. These molecules are comparatively easy to design: many dozens have been made, vs. the few proteins designed *de novo*. Their polycyclic structures, however, usually present synthetic challenges greater than those of peptides of similar molecular weight. Their molecular masses typically fall in the 200 to 1200 dalton range, vs. ~2000 daltons for commercially available custom peptides and 5000 to 15000 daltons for experimental products.

These two approaches, growing out of protein chemistry and more general chemical synthesis, represent extremes on a spectrum. In the first, synthesis is relatively straightforward, but design is difficult; in the second, design is relatively straightforward, but synthesis is difficult. Intermediate approaches may avoid the worst problems of each, by exploiting peptide-style synthetic strategies to construct molecular objects more rigid, stable, and designable than natural proteins. In examining the potential of this approach, fundamental principles of design are relevant.

Basic Design Principles

Designers exploit *modularity*: In his classic essay, "*The Architecture of Complexity*," Herbert Simon notes the utility of modular structure in understanding, designing, and building complex systems. Organizing a system as an assembly of relatively distinct, independent building blocks enables the use of divide-and-conquer strategies in design and construction. Early successes in protein design have used a helices as modular units in design and development.

Designers exploit *safety margins*: Evolutionary considerations suggest that evolved proteins can be expected to have minimal stability while designed proteins can have a greater margin of stability; recent experience shows that protein designers can indeed build molecules of increased stability. Safety margins in stability will improve the odds of success in designing molecular functions.

Finally, designers exploit *design freedom*: In design, unlike the analytical sciences, one is free to seek components having the best available combinations of stability, predictability, manufacturability, and performance. Design freedom in protein engineering enables the successful pursuit of modularity and safety margins, as has been demonstrated.

Alternatives to Standard Proteins

Structures based on sequence-specific polymers have advantages worth keeping. To assemble complex macromolecular objects from small molecules, linkages must be introduced in stages. Ease of synthesis favours starting by building linear, polymeric structures. Established synthetic techniques, the entropic advantages of rotationally rigid bonds, and the extensive knowledge resulting from protein science all favour further exploitation of peptide chemistry to build those polymers. (Note, however, that a similar case could be made for exploiting nucleic acid chemistry.)

Noncovalent linkages are well suited for the subsequent steps: they can form under conditions that preserve existing covalent structures; they can link complementary, polyatomic surfaces, enabling specific binding; and they can allow equilibration, breaking and reforming linkages, thereby enabling the system to "search" for a cooperative binding configuration planned by the designer. (Disulfide bonds also allow equilibration under mild conditions and are of similar utility.) Accordingly, proteinlike structures have engineering advantages quite apart from their biological interest.

Within this framework, design freedom can be expanded by choosing from a broader range of parts and chain topologies. It may be that linear sequences of the 20 genetically encoded amino acids of *standard proteins* will provide adequate design freedom, eventually making *de novo* protein design practical and routine. Nonetheless, it may be useful to examine the advantages and disadvantages of *augmented proteins* made using (1) a set of monomers augmented by inclusion of nonstandard amino acids and (2) a set of chain topologies augmented by branched and cyclic structures. Augmented structures have repeatedly proven their utility in peptide design, and nonstandard amino acids have appeared in proposals and experiments in protein design. Nature's use of nonstandard amino acids, phosphorylation, glycosylation, and so forth (via enzymatic modification of proteins) also indicates the utility of a broader set of building blocks and structures than the 20 genetically encoded amino acids.

The following sections build on current understanding of protein folding and stability to explore the utility of augmented proteins for designing stable folded structures. Many of these remarks apply equally well to nonpeptide structures, which would further expand design freedom but are not explicitly considered here.

Strategies for Stabilizing Specific Folds

Achieving a stable fold entails destabilizing (or excluding) both unfolded and misfolded states. Achieving a stable and *predictable* fold requires that the designed fold be clearly stable and that alternative folds be clearly improbable; this requires a margin of safety that exceeds modeling uncertainties. Both of these objectives can be pursued by (1) minimizing conformational entropy in the unfolded chain, and by (2) maximizing the strength and specificity of interactions favouring a chosen folded state.

Multiple stabilizing strategies can be applied, but quantifying their value in a design context raises questions regarding the choice of reference state. The *stability* of a protein can be defined as $-\Delta G_f$, the negative of the change in Gibbs free energy of folding. The change in stability resulting from a given residue substitution is then $-\Delta(\Delta G_f)_{sub}$. Employing a stabilization tactic in a design, however, will typically involve not a single substitution, but a coordinated set of changes. Consider two protein molecules, A and B, having grossly similar dimensions and properties. Assume that each has been designed for maximum stability, subject to similar functional constraints, but that use of a tactic X (e.g., use of a dialkylglycine residue, a cross-link, etc.) is allowed only in the design of B, which may differ from A in any manner needed to exploit X effectively. Tactic X should be considered stabilizing if B can *usually* be made more stable than A. More generally, one can define the *expected stabilization* resulting from the application of a tactic X as the average change in $-\Delta G_f$, resulting from the comprehensive redesign of typical molecules to take advantage of X; this can be represented as $-\Delta(\Delta G_f)_{design}$. The magnitude of $-\Delta(\Delta G_f)_{design}$ cannot be measured directly without an enormous amount of design and experimentation. Since good designs will avoid unfavourable interactions in both A and B, however, an estimate of the value of $-\Delta(\Delta G_f)_{design}$ is the value of $-\Delta(\Delta G_f)_{sub}$ that is expected when a single-site substitution (cyclization, etc.) introduces no adverse interactions, such as strain or steric crowding. This will be referred to as a *compatible* modification.

Backbone rigidity

Decreasing the conformational freedom of the backbone decreases the entropy of unfolding, thus stabilizing a compatible folded state. In accord with theoretical predictions, the compatible replacement of Gly with Ala (or, it is expected, with any amino acid having a β carbon) or of Ala with Pro each increase stability by ~7 maJ; placing Pro in an α helix, however, would illustrate an *incompatible* replacement. In augmented proteins, various cyclic amino and imino acids could be used to constrain conformations in a similar manner, but with varied geometries.

Peptide researchers have exploited nonstandard amino acids and characterized their conformational effects. In particular, dialkylglycine residues impose severe restrictions on ϕ and ψ angles; relative to Ala, they should yield a $-\Delta(\Delta G_f)_{design}$ on the order of ~7 maJ per use.

R-group rigidity

Folding flexible R-groups into a tightly packed, hydrophobic protein core reduces their conformational entropy. This adverse effect can be reduced by making greater use of relatively rigid R-group structures, such as the rings of phenylalanine, tryptophan, or nonstandard amino acids. On entropic grounds, each elimination of a rotational degree of freedom from an aliphatic chain should yield a $-\Delta(\Delta G_f)_{design}$ equaling ~4 maJ.

Like other constraints compatible with a planned fold, increased R-group rigidity inhibits unplanned folds. The potential for unplanned folding is suggested by the work of Ponder and Richards on defining sets of residue sequences and R-group conformations compatible with a given backbone conformation. A geometric test for sequences of core residues compatible with the fold of rubredoxin yielded 44 sequences, but 236 sets of compatible conformations-that is, an average of over 5 plausible core packings per sequence. Allowing small variations in backbone conformation would presumably have enlarged this number. Both lamprey hemoglobin and crambin have internal residues with alternative conformations.

Greater use of rigid R-groups makes alternative packings less likely, increasing the specificity of packing and folding. However, it will also increase the difficulty of finding a sequence that has even one acceptable packing.

Steric diversity

More rigid R-groups provide fewer shapes per structure, but this can be countered by providing more structures. How many shapes are needed to make tight packings possible? The nonpolar amino acids Ala, Val, Leu, Ile, Pro, Phe, and Trp assume 18 common rotameric conformations in proteins; in a sample protein population they make up 37% of the residues, and a larger fraction of the interior residues. This suggests that use of a few tens of relatively rigid R-groups can enable the design of packings having a large fraction of rigid structure.

Natural protein interiors have a mean volumetric packing density of about .74 with local variations from $<.6$ to $>.85$. A cavity the size of a methylene group decreases the stability of the folded state by about 8 maJ, and various unfavourable interactions in the folded state (e.g., adverse van der Waals contacts) reduce stability by an estimated 14 maJ per residue. Expanding design freedom by including D-amino acids and nonstandard R-groups should increase achievable packing densities while reducing adverse van der Waals contacts-a wider choice of pieces will help designers find better-fitting patterns.

Pairwise matching

An unworkable—yet instructive—way to think of protein folding is to regard each interior R-group as binding to a receptor site formed by the rest of the fold. In this entirely nonmodular picture of cooperative folding, every part fits because of the way the rest fits, offering no point of departure for design and analysis. In standard proteins, regularities at the level of secondary structure provide a partial escape from this quandary.

In augmented proteins, a large R-group could provide a preformed, fold-independent binding site for another R-group, yielding modularity at the level of monomer pairs. Supramolecular complexes and nucleic-acid base pairing in folded RNA and in DNA junctions provide models for such pairwise specificity. Even weak pairwise binding could aid folding specificity, while stronger interactions could strongly direct folding by forming complexes stable enough to mimic covalent cross-links.

Cyclic and branched backbones

Fold-compatible cross-links decrease the entropy of unfolding, stabilizing the folded state (by about 2.8 maJ per residue in a 10-residue loop at room temperature). Engineered disulfide cross-links have been used to stabilize natural proteins. Since looped structures (closed, for example, by amide bonds between R-groups) can be made using solution chemistry and then treated as monomers during solid-phase synthesis, augmented proteins can incorporate cyclic backbone segments.

Studies of omega loops show that small looped structures (6 to 16 residues) can be globular and densely packed. Analogous cyclized structures can be regarded as small domains or large monomers, serving as modular elements in design and folding. These elements can be large enough to exhibit pairwise selectivity and binding, strongly guiding the folding process.

Branched structures decrease the entropy of unfolding through excluded-volume effects. This augmentation strategy has brought striking success in constructing stable tertiary structures, including a 73-residue molecule with enzymatic activity.

Staged thiol pairing

Many standard proteins form cycles postsynthetically, through oxidation of pairs of cysteine thiol groups to form disulfide bonds. Cross-links exposed to a suitable mix of thiols and disulfides in solution undergo rapid thiol-disulfide exchange through S_N2 displacement; this can anneal the disulfide bond system toward a pairing pattern of minimal free energy. In augmented proteins, a wider choice of thiol R-groups can broaden the range of disulfide cross-link geometries. Further, bulky monomers (or cyclic backbone segments) can provide local steric constraints to guide cross-link formation.

Formation of disulfides under conditions of increasing oxidation and (initially) high pH can favour pairing of thiols in stages corresponding to their pK_a values, from highest to lowest, making pairing more specific. Data from studies of the kinetics and equilibria of thiol-disulfide exchange indicate that, although equilibration among disulfides would destroy pK_a-specific pairing, lower pK_a thiols can be paired and equilibrated among themselves rapidly enough to limit destructive equilibration with established, less-labile disulfides formed from thiols of substantially higher pK_a. Preliminary Monte Carlo simulations of staged thiol pairing in this approximation indicate that molecules having solely within-class pairing (for 12 to 18 thiols in three pK_a classes) can be obtained with yields of tens of percent.

In this sequential annealing process, early-forming bonds can sterically constrain later bond formation (as in standard proteins). This mechanism can hinder incorrect pairings within a set of thiols of a particular pK_a, enabling the formation of multiple specific cross-links per stage. The chief disadvantage of disulfides formed from low-pK_a thiols is extreme sensitivity to alkaline degradation.

Table 4.1. Model thiol compounds and their pK_a values in aqueous solution.

Thiol	*pK_a*
Thiobenzoic acid	2.48
Thiolacetic acid	3.5
2-Nitro-5-mercaptobenzoic acid	4.53
4-Mercaptobenzenesulfonic acid	5.73
Thiophenol	6.5
Methyl mercaptoacetate	7.68
Cysteine	8.30
1-Thioglycerol	9.51
Ethanethiol	10.61
2-Methyl-2-butanethiol	11.35

Metal complexation

Suitable R-groups can form constraining cross-links when metal ions are introduced into solution. In standard proteins, these constraints can force folding of structures as small as 3D-residues (e.g., zinc fingers). Engineering of metal binding sites has increased the stability of subtilisin. Work in supra molecular chemistry suggests that single

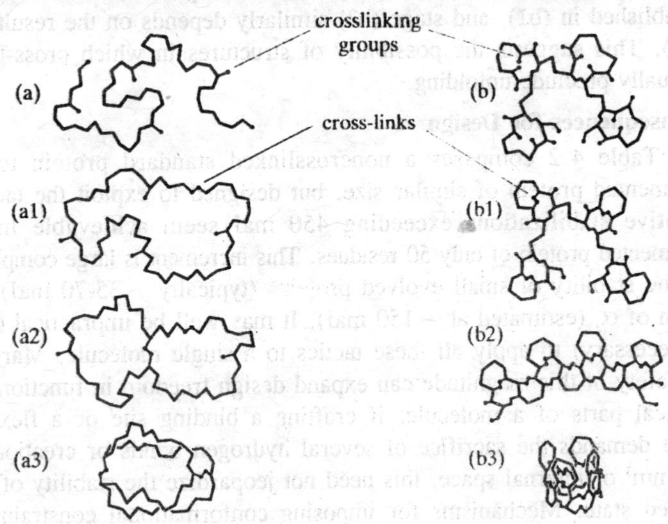

Fig. 4.1. Staged crosslinking and cyclic backbone segments.

R-groups (e.g., based on α,α' bipyridine) can serve as preformed halves of metal binding sites, undergoing pairwise matching when metal ions are added to a solution. As in staged thiol pairing, the time at which crosslinking occurs can be controlled by manipulation of solution composition. Thus, metal complexation can add a further stage to the strategy of forming cross-links in locations determined by constraints established by previous cross-links. Metal binding can also drive the quaternary assembly of separately folded structures; disulfide formation can do likewise.

Series (a) shows in schematic form how six crosslinking groups in three reactivity classes (based on thiol pK_a values or on distinct binding specificities) might be used to constrain conformation through staged crosslinking. In stage (al), the molecule is cyclized; in (a2) and (a3) the initial loop is further crosslinked. A theory of entropy for multiply-looped chains indicates that a fold based on (a3) would be stabilized by roughly 105 maJ relative to the uncrosslinked structure (at 30°C). More speculatively, series (b) suggests an extensive application of staged crosslinking in which multiple crosslinking groups pair at each stage and selectivity of pairing within a class (if achievable) is dependent on constraints resulting from prior structure. In this sketch, a chain is synthesized with cyclic backbone segments designed to sterically and entropically favour the pattern of pairing shown in stage (b1). Selectivity in stage (b2) depends on adequate steric constraints established in (b1), and stage (b3) similarly depends on the results of (b2). This suggests the possibility of structures in which cross-links virtually preclude unfolding.

Consequences for Design

Table 4.2 compares a noncrosslinked standard protein to an augmented protein of similar size, but designed to exploit the tactics relative stabilizations exceeding 450 maJ seem achievable in an augmented protein of only 50 residues. This increment is large compared to the stability of small evolved proteins (typically ~35-70 maJ), or even of α_4 (estimated at ~150 maJ). It may well be impractical (and unnecessary) to apply all these tactics to a single molecule. Margins of safety of this magnitude can expand design freedom in functionally critical parts of a molecule: if crafting a binding site or a flexible joint demands the sacrifice of several hydrogen bonds or creation of 0.1 nm³ of internal space, this need not jeopardize the stability of the folded state. Mechanisms for imposing conformational constraints—including backbone rigidity, R-group rigidity, cyclic segments, and staged

Table 4.2. Stability increment expected for a 50-residue augmented protein.

Mechanism	*Stability improvement (maJ)*
Increased backbone rigidity	115
Increased R-group rigidity	50
Improved packing density	70
Reduction in bad van der Waals contacts	?
Cyclic backbone segments	110
Disulfide cross-links, metal complexation	110

thiol- and metal-based crosslinking-will not only stabilize the desired fold, but will destabilize (or simply block) an indefinitely large set of alternative folds.

The destabilization of unwanted alternatives can be quantified: since ~275 maJ of the stability increment results from decreases in conformational entropy imposed before folding begins, the number of energetically accessible alternative configurations (including both unfolded, misfolded, and misaggregated states) is reduced to $\sim 10^{-29}$ of the unconstrained value. By making results more predictable, this should increase the success rate in the trial-and-error process of designing macromolecular objects. Several groups have developed software to aid protein design; these systems appear adaptable to augmented proteins. Further, stability achieved through rigidity reduces the sensitivity of designs to the errors in molecular mechanics models; these errors pose a serious problem where natural proteins are concerned.

This discussion has focused on backbone and hydrophobic core structures, but surface properties largely determine molecular behaviour. These, too, are targets for design. For example, solubility—though aided by structures that favour the folded state and thus disfavour alternative, aggregated states—can be further increased by the generous use of polar and charged R-groups on the molecular surface; these can include groups bearing polar or multiply-charged polymeric chains (large, solubilizing side chains that can be cleaved and discarded after a molecular object has bound to a larger assembly may be of particular value). Broadened residue options will facilitate the design of surface geometry, electrostatics, reactivity, and so forth.

Trade-offs and Applications

By expanding design freedom, the use of augmented proteins can aid the design of stable, predictable, functional molecules. Several of the tactics proposed here for augmented proteins may prove difficult to implement, but avoidance of those that prove too difficult is a proper use of design freedom. Commercial solid-phase synthesis of standard peptides now offers products with costs of ~$10 per milligram for $1000 orders of crude oligopeptides in the 10-mer to 20-mer range. Prices climb steeply with chain length and purity; they fall steeply with increased quantity. Lengths near the 50-mer range are approaching routine practice.

Feasible tactics should yield stable folds in chains of readily accessible length. Small size aids characterization (without crystallization) via solution nuclear magnetic resonance techniques; these have been extended to structures several times as large as those contemplated here. Larger molecular systems can then be designed as quaternary structures self-assembled from these building blocks; a review of relevant principles appears in Lindsey (1991), and a more general survey appears in Whitesides et al. (1991). Various high-value molecular products might be worth making, despite the cost of chemical synthesis.

Macromolecular Objects via Mechanosynthesis

Motivation

The atomic force microscope (AFM also termed a scanning force microscope) enables surfaces to be positioned relative to one another in a solution environment with ~0.01 nm positional precision and tip compressive loads as low as 0.01 nN. AFMs are most often used to form images of surfaces by probing mechanical interactions at many points. This section suggests how AFM mechanisms could be exploited to enable solution-phase mechanosynthesis; instruments of the sort described also have more immediate applications in improving AFM imaging.

The lack of reproducible, well-characterized, atomically sharp tips in standard AFM systems causes difficulties in image resolution, interpretation, and reproducibility. It has been proposed that AFMs with molecular tips could both alleviate these problems and provide an approach to achieving positional control of chemical synthesis. Single molecular tips, however, present problems of fabrication, yield, and damage during use.

Molecular tip arrays as proposed here can permit screening and interchange of tips during operation, reducing sensitivity to yield and damage. They can enable sequential use of differing tips to image a sample and to perform solution-phase mechanosynthesis. A penalty of this flexibility is limitation to small substrates. R. Merkle has proposed an analogous approach based on scanning tunneling microscope mechanisms working with adsorbed reagent molecules in vacuum.

Tip-array Geometry and Forces

Tip-array geometry

On a large scale, a *primary bead* can be viewed as an AFM tip, imaging an array of *flat-side beads*. On a small scale, a single molecular tip on the flat-side *secondary bead* images the primary bead. The terms *substrate* and *tip* are thus ambiguous, but under the conditions of greatest interest, the primary bead acts as the substrate and a molecule on the secondary bead acts as the tip. Key geometrical parameters are defined. Reasonable dimensions for a device with this geometry are $r_1 \approx 500$ nm, $r_2 \approx 50$ nm, $h \approx 4$ nm, and $\delta \approx 6$ nm. Commercial AFMs enable operation in a fluid medium with atomic resolution and scan ranges of 25 μ; they seem suitable as basic mechanisms. The following assumes that the fluid medium is an aqueous solution, but anhydrous solvents are feasible and sometimes desirable.

If bead-bead forces and stiffnesses are small relative to the required imaging forces and stiffnesses, they will permit imaging based on interactions between the tip and the target structure. To limit bead-bead interactions, a tip must have a height h greater than some

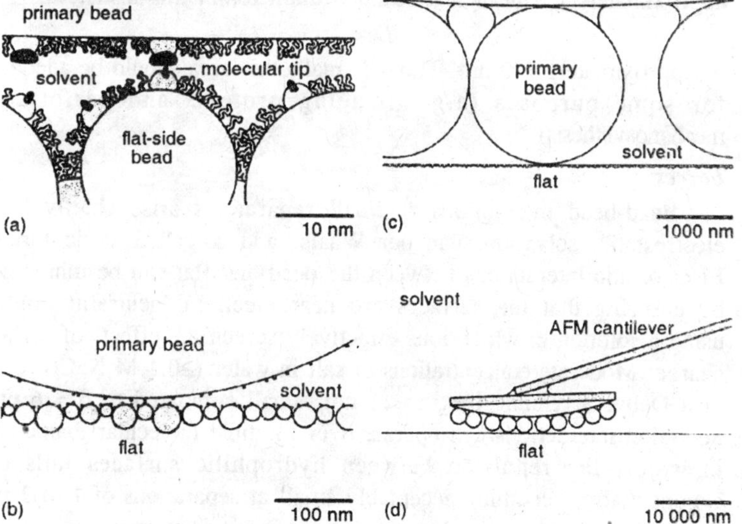

Fig. 4.2. Views of a tip-array geometry using tip support structures and multiple substrate beads on an AFM cantilever.

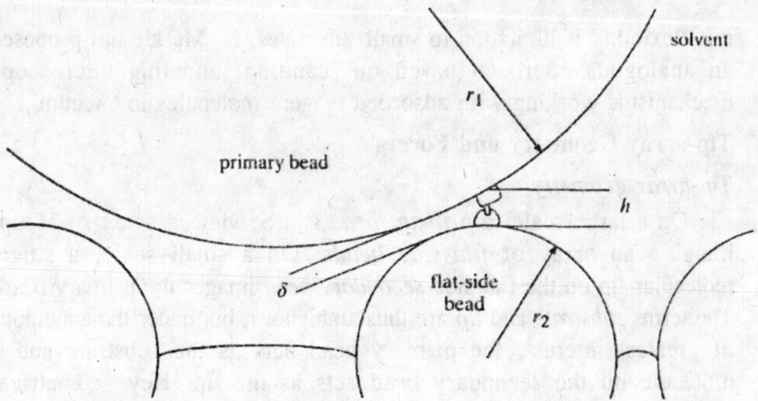

Fig. 4.3. Sketch illustrating the relationships among the bead, flat, and molecular tips, including basic geometric parameters.

minimum imaging separation δ. For typical geometries with $r_1 > r_2 \gg (h - \delta)$, a region of diameter

$$d \approx \sqrt{8(2h-\delta)(r_1+r_2)r_1/r_2} \qquad \text{...(1)}$$

on the primary bead can be scanned to image or modify target structures, so long as other flat-side beads do not interfere. With $r_1 = 500$ nm, $r_2 = 50$ nm, and $(2h - \delta) = 2$ nm, $d \approx 300$ nm. Since the centers of the flat-side beads are expected to be separated by approximately r_2, the actual scanned diameter in this instance is

$$D \approx 2r_2 \qquad \text{...(2)}$$

or approximately 100 nm. Though small, this area should be adequate for some purposes (e.g., imaging proteins and performing mechanosynthesis).

Forces

Bead-bead interactions at small separations arise chiefly from electrostatic, solvation, van der Waals, and so-called steric forces. Electrostatic interactions between the bead and flat can be minimized by ensuring that the surfaces are near electrical neutrality, or by using a solution in which ions effectively screen the effects of surface charge. Moderate concentrations of salt in water (≥0.1 M NaCl) result in a Debye screening length < 1 nm, small compared to δ ≈ 6 nm. Solvation interactions can operate over multiple molecular diameters: in water, the repulsion between hydrophilic surfaces falls off exponentially, becoming acceptably small at separations of 1 to 3 nm (hydrophobic forces would cause strong adhesion between surfaces and must be avoided). Oscillating solvation forces associated with molecular size effects become small at separations greater than ~2 nm in water, and are reduced by surface roughness and mixed-solvent systems. With rough bead surfaces and $\delta \approx 6$ nm, oscillating forces should not be observed.

Bead-flat van der Waals interactions can be described using continuum models based on the Hamaker constant. The results are thought to give a reasonably accurate picture of the forces and their dependence on geometrical parameters. One family of approaches to bead fabrication would use polystyrene-like materials; for polystyrene interacting through water, the Hamaker constant $A = 1.3 \times 10^{-20}$. With the given geometrical parameters and an intersurface separation $s = \delta = 6$ nm the force F is attractive and ≈ -0.003 nN. This is smaller than the minimum tip contact force reported in AFM work, 0.01 nN, and is associated with a stiffness of only ~-0.001 N/m, which is negligible in comparison to the stiffness of a typical AFM cantilever or of an interaction between two molecules in contact. Accordingly, this force should cause little difficulty.

Previous papers have described less-symmetrical structures, with support molecules attached on a single side to avoid a requirement for two distinct attachment technologies. In those geometries, intersurface separations are smaller, requiring material and solvent choices that minimize the net Hamaker constant. The present approach has greater symmetry, enabling a single attachment technology to mount both target-supporting and tip-supporting molecules. The resulting increase in inter-surface separation relaxes constraints on the Hamaker constant, and hence on the choice of materials and solvents.

The use of short solvophilic chains (e.g., polyoxyethlyene) on the bead surfaces to provide a short-range repulsive force caused chiefly by entropic effects. This mechanism is termed *steric stabilization* and is widely used to prevent the coagulation of colloids. Attachment of these chains renders the bead surfaces hydrophilic and inhibits the bead-bead adhesion that would otherwise occur though short-range hydrophobic and van der Waals attractions.

Multiple beads

Larger-scale geometrical features of a system that permits primary beads to be interchanged during operation by tilting the flat relative to the cantilever. Except for molecular structures at the finest scales, and platelike structures at the largest scales, each of the shapes illustrated can be generated by the solidification either of a droplet or of a fluid forming a meniscus between two objects. With the same exceptions, all close contacts occur between spherical surfaces and are insensitive to angular misalignment. Accordingly, the most severe fabrication challenge appears to be that of attaching suitable molecules to beads, and doing so with adequate stiffness.

Molecular Tips and Supports in AFM

Protein-ligand tips

Potential tip structures include proteins or particles bearing small adsorbed molecules. A particularly versatile approach exploits the capabilities of both organic synthesis and biotechnology by using synthetic ligands as tips and protein molecules as supporting structures. Many natural proteins bind partially exposed ligands. Ligand analogs could be synthesized with protruding moieties having steric properties suiting them for use as AFM tips. Extensions of monoclonal antibody technology can rapidly generate proteins able to bind almost any selected small molecule. Single-chain Fv proteins (combining antibody V_L and V_H sequences) are compact, ~3 nm in height, and lack hinge regions. Use of these antibody-derived proteins can allow broad freedom in ligand design, if their stability is (or can be made) adequate for use in an AFM mechanism.

Stiffness

To be used as AFM tip-supports and as imaging targets, proteins and protein-ligand complexes must have adequate mechanical stiffness. In protein crystals (which have been used as models for bound protein-protein complexes), individual atoms in the protein interior typically undergo a 0.03 to 0.05 nm root-mean-square displacement caused by thermal vibration. The atomic-displacement stiffness, k_s, can be derived from the root mean square displacement via the relationship for a thermally excited harmonic oscillator, which implies $k_s \approx 1.6$ to 4.6 N/m for typical atomic displacements in the interior (k_s for a side chain on a surface often is far lower). Proteins in crystals typically are anchored to neighbours by a few side chain contacts; comparably stiff attachment of proteins to surfaces seems achievable.

Rigid, polycyclic structures of substantial size can be made by organic synthesis, and their internal stiffnesses can exceed those of proteins. Such ligands can be anchored relative to the protein by numerous van der Waals interactions of substantial stiffness, yielding ligand atomic-displacement stiffnesses toward or beyond the upper end of the range characteristic of proteins: the following sections will assume a value of 3 N/m. The stiffness of the interaction in imaging a protein by a protein/ligand complex can thus be ≥1 N/m; AFM cantilevers with k_s < 1 N/m should provide good responsiveness.

Stability of proteins under compressive loads

Applied forces can destabilize protein folding and ligand binding. The free energy required for unfolding or unbinding typically is ~50 to 100 maJ; at the larger energy, the unfolding half-life is on the order of 1000 years (assuming a 1 s folding time; typical values range from 0.1 to 1000 s). From a kinetic perspective, the destabilizing energy increment resulting from a force equals the work it performs as the molecule moves from equilibrium to a transition state for unfolding (where the location of the transition state is affected by the force); estimating this energy requires an estimate of the atomic displacements associated with such a transition state.

In a linear elastic system, the strain energy V_s is related to a displacement Δx by

$$\Delta x = (2V_s / k_s)^{1/2} \qquad \text{...(3)}$$

Large V_s and low k_s yield a large (adverse) value of Δx. For V_s = 100 maJ and k_s = 1.6 N/m, Δx = 0.35 nm. Localized displacements of this magnitude (an atomic diameter) might independently be expected to disrupt the tight core packing necessary for stability. Over this displacement, 0.1 nN performs 35 maJ of work, which is small compared to the energetic differences between more and less stable proteins. Thus, on kinetic stability grounds, this estimate suggests that 0.1 nN forces should be compatible with imaging of proteins of moderate stability (occasionally interrupted by denaturation), and compatible with the use of tips based on proteins of good stability incorporating well-bound ligands.

Considering only stiffness and acceptable elastic deformations (taken to be 0.01 nm), and calculating from a continuum model, a maximal tip force of 0.01 nN had been suggested for protein imaging. Note, however, that with $k_s \approx 1.6$ N/m (a low value from experimental data), a 0.1 nN force would yield Δx = 0.06 nm and $V_s < kT_{300}$. These values are compatible with imaging yielding useful structural data.

In AFM systems of standard geometry, repulsive interatomic tip forces (in the presence of net long-range attraction) have been reduced to low values. It had been suggested that forces as low as 0.01 nN should be within reach for systems operating in solution, and this was subsequently achieved. Comparably low forces should likewise be possible in multiple-tip systems, providing a substantial margin of safety when using proteins as tip supports or as imaging targets.

Attachment of Supporting Molecules

The immobilization of molecules (and proteins in particular) is a well-established technique in biotechnology, often used to retain molecules in a vessel as fluid flows past. Immobilization commonly relies on the formation of cross-links between reactive moieties on a protein and on a surface. Molecular tip arrays, however, will require more than immobilization against macroscopic displacement: they will require relatively stiff attachment of molecules to a surface.

Immobilization requires only a molecular tether; stiff attachment requires three (or more) points of contact providing tripodlike stability. The stiffness of a tip can be no greater than the stiffness of the interface attaching it to a surface. A stiffness of ~1.6 N/m combined with a tip contact load of ~0.01 nN can produce an effective concentration of ~145 nm^{-3} for a tip moiety relative to a point on a substrate, in the presence of AFM jitter and thermal noise. This stiffness is compatible with the mechanical properties of proteins, assuming good attachment. It also suffices for strong positional control of chemical reactions.

As suggested already that strategies for achieving stiff attachment of proteins to surfaces, several of which exploit the engineering of protein-surface side chains to provide specific binding interactions. One family of approaches would attach proteins to surfaces directly, via the bonding of side chains (e.g., the thiols of cysteine) or chemically-modified side chains. Another family of approaches would use techniques akin to the freeze-cleave-etch methods used to prepare specimens for electron microscopy. A final family of approaches discussed there would link proteins to molecules embedded in fluid Langmuir-Blodgett films, which could then be polymerized to increase their stability; this has the advantage of forming (in effect) a binding site for the protein, thereby increasing its stability. A more recent approach, suggested by the Langmuir-Blodgett film approach combined with ideas suggested by work on imprinting small molecules in polymers would link proteins to surfactant molecules embedded in a droplet that is later polymerized to form a bead. In light of these possibilities, suitable attachment techniques appear feasible.

Imaging with Molecular Tips

On a large scale, scanning the primary bead provides a low-resolution image of the array of flat-side beads. In regions where the primary bead and a secondary bead interact through steric repulsion forces between the polymers, the image is expected to be essentially featureless. In regions where a molecular tip on the secondary bead interacts with a target structure on the primary bead, the image is expected to show structural features at high resolution. Systematic probing in these regions (moving in and out by several nanometers at a series of points on a grid) can generate a map of force vs. distance for the interaction between tip and target. A surface in that map corresponding to a large, constant stiffness should in many instances resemble an atomic-resolution topographic map.

For a system with a $\sim 25\ \mu$ scan range and $d \approx 100$ nm, the number of immediately available tips can be $\sim 10^5$. A tip array can advantageously include tips that have diverse structures and orientations.

On an isotropic substrate, supporting molecules (and their bound tips) will be attached at diverse azimuthal angles. Bead curvature and structural irregularities will increase the diversity of orientations in the other two rotational degrees of freedom.

Tips can be polar, nonpolar, hydrogen bonding, positively charged, or negatively charged. Information gained by imaging a single molecule with tips of multiple types may help fulfill the goal of determining folded protein structures through AFM and computational modeling. It may be that features can be reproducibly mapped on the surface of a protein, yet fail to indicate which residues occupy which surface locations. If so, then imaging and comparing a series of proteins (prepared via site-directed mutagenesis) that differ only by the substitution of a distinctive residue at a known position in the sequence could reveal much about the structure of the folded state.

Tips with specific biochemical affinities (e.g., enzyme substrate analogues, candidate drug molecule analogs, etc.) can yield data of special biological interest. A tip can be moved along a trajectory aligned with the flexure direction of the AFM cantilever, carrying it from solution to (for example) a protein surface and back again. Integrating the force curve (with suitable corrections for bead-bead interactions) then yields an intermolecular mean-force potential, that is, a high-resolution spatial map of the free energy of the intermolecular interaction. For probes with tip structures of biological interest, this information can be of direct biological relevance regardless of the apparent image resolution.

To facilitate topographical imaging, organic synthesis can be used to prepare ligands with stiff tips of considerable geometric acuity. For example, molecular mechanics calculations (MM2/CSC) indicate that the H atom of R—C≡C—H has a transverse bending stiffness of ~2 N/m relative to an sp^3 carbon in the R-group (e.g., in a polycyclic ligand structure); the longitudinal k_s is ≥100 N/m. The effective tip radius for such a probe is ~0.13 nm.

Tips made by current microfabrication techniques have irregular structures with characteristic radii estimated to be ~10 nm. The molecular tips proposed here can have well-characterized structures of diverse types, attached in diverse orientations. Imaging specimens of known structure using multiple tips should permit discrimination and mapping of the tip types and orientations in a particular tip array. To take a simple example, positive, negative, and neutral tips will have distinct, contrasting responses to a bound charge.

Chemists frequently prepare derivatives of molecules to identify the original structure. Ligands used as molecular tips can include chemically reactive moieties that can be used to map sites of differing reactivity. This mode of operation will usually require prolonged dwell times at a candidate site, as opposed to steady scanning or rapid probing.

Solution-phase Mechanosynthesis

Organic synthesis today offers powerful techniques for building molecular structures. Use of maneuverable molecular tips can add a fundamental novelty: flexible, positional control of sequences of synthetic steps based on control of local effective concentration. Receptors that bind reagent ligands can serve as tool holders for mechanosynthesis.

Molecular tip array systems enable approaches substantially different from those previously proposed: Tip arrays and antibody technologies will enable the use of distinct receptors for reagents of different types, permitting a series of reactions without cycling the composition of the solution. With rapid, spontaneous dissociation no longer necessary for interchange, reagent ligands can be bound tightly; this enables larger effective concentration ratios and hence better site specificity.

Effective concentrations

Reagent reaction rates are by definition proportional to reagent effective concentrations. All else being equal, the effective concentration of a reagent is proportional to its spatial probability density. Modeling a reagent ligand as a thermally excited object, subject to a force F pressing it against a barrier, to constraint by a transverse stiffness k_s, and to an additional Gaussian jitter in the AFM mechanism with a root mean square amplitude = σ_{AFM}, yields a probability density having a peak local concentration (m^{-3})

$$c_{local} = \frac{F}{2\pi kT}\left(\frac{kT}{k_s} + \sigma_{AFM}^2\right)^{-1} \quad \ldots(4)$$

Neglecting energetic and orientational effects (which can, when favourable, greatly increase the effective concentration), and provided that k_s is substantially less than the transverse stiffness in the reaction transition state (as is the case here), local concentration corresponds to effective concentration. Displacements of well under 0.01 nm are now routinely measurable in stationary tips, but jitter from ambient vibration is of this order. Assuming σ_{AFM} = 0.01 nm, T = 300 K, F = 0.01 nN, and k_s = 1.5 N/m (between two structures with stiffnesses of 3 N/m) yields c_{local} = 135 nm^{-3} (~225 M) at 300 K. For comparison, mobile surface-residue thiol groups in proteins can exhibit effective concentrations exceeding 60 nm^{-3}; interior residues often exhibit much larger values.

Rate constants for ligand binding commonly exceed 10^8 nm^3 s^{-1}. Antibody technologies commonly yield proteins that bind a chosen ligand with equilibrium dissociation constants on the order of 10^{-9} nm^{-3}. On exposure to a solution with a 10^{-6} nm^{-3} ligand concentration, reagent receptors with these properties will become occupied in ~10^{-2} s, and at equilibrium will remain occupied with probability ~.999 (with these parameters, the mean waiting time for dissociation to occur is ~10 s). Application of mechanical load to a tip tends to increase its equilibrium probability of dissociation while decreasing its rate of dissociation; this behaviour is compatible with the tip applications described here.

Under the conditions just specified, the peak enhancement in effective concentration of a reactive moiety on a molecular tip relative to the background solution concentration is ~(100 nm^{-3})/(10^{-6} nm^{-3}) = 10^8. This localized, positionable enhancement in effective concentration can direct site-specific reactions on molecular structures attached to the bead. The transverse probability density distribution is Gaussian; with these parameters, the enhancement falls to background levels at a radius of ~0.3 nm, or about one atomic diameter. Thus, this mechanism for positional control can produce reaction rate differentials on the order of 10^8 between otherwise chemically equivalent sites separated by two to three bond lengths.

Anticipated synthetic capabilities

The choice of reagent ligands is constrained by requirements for chemical compatibility. Protein molecules have been suggested as reagent receptors; they have reactive functional groups and ordinarily exist in an aqueous medium. Modifications (including immobilization itself) may help stabilize proteins, permitting their use in polar, nonaqueous media.

Reagent ligands also must not react with one another at substantial rates in solution, unless the resulting by-products are harmless to the process. Low concentrations reduce this problem, and a system that steadily flushes the AFM cell with freshly mixed reagent solution (ensuring a low residence time) can further reduce unwanted reactions. If the residence time is ~ 100 s, the effective-concentration ratio is $\sim 10^8$, and a typical mechanosynthetic reaction time is ~ 1 s (at an effective concentration of 100 nm^{-3}), then the concentration of side-reaction products will be $\sim 10^{-12}$ nm^{-3}, or $\sim 10^{-6}$ that of the reagents.

A mature AFM-based mechanosynthetic technology could potentially perform syntheses involving $> 10^5$ sequential steps in the presence of ~ 100 chemically equivalent sites with only a small probability of failure resulting from reactions with uncontrolled reagent molecules in solution. In an operating system, each reaction step removes a reagent ligand from a receptor that is promptly recharged by a molecule from the solution. Each reaction step can be tested for success by AFM imaging, to ensure completion. The extension of tip-array technology to encompass solution-phase mechanosynthesis will require a multidisciplinary collaboration with synthetic organic chemistry playing a prominent role. A major objective is the identification of building blocks that are (1) accessible via conventional solution-phase synthesis, and (2) adequate for use in building molecular machine systems, given positional control of the expected reliability; initial studies have been undertaken by M. Krummenacker.

Summary

Analysis suggests that AFMs with arrays of molecular tips may have substantial advantages over conventional AFM systems for the intensive study of molecular-scale specimens, including protein molecules. The chief sacrifice is a drastic reduction of the effective substrate area, which is limited to a fraction of the bead diameter. The opportunity to select among $\sim 10^6$ tips, guided by imaging results and without interrupting system operation, sidesteps the requirement for good tip yield and durability imposed by single-tip systems. Selection of tips with desirable mechanical and orientational properties from a population comprising well-defined molecular structures will enable more thorough and unambiguous characterization of samples. Use of reactive molecules as tips will enable nanofabrication by positional control of chemical synthesis. The estimated reliability is sufficient to enable $>10^5$ sequential steps, enabling the construction of molecular objects $\sim 10^3$ times larger than typical protein molecules and existing products of chemical synthesis.

CONCLUSIONS

Present technology can be extended to build macromolecular objects and systems in any of several ways:

1. Biotechnology (and protein engineering in particular) can be extended to make ~ 300 monomer molecular objects capable of forming larger structures of indefinite size and complexity by Brownian assembly. Problems with design are the chief obstacles to this path.
2. Solution-phase synthesis can be used to make ~ 50 monomer molecular objects with greater cost per unit mass, but with lessened design difficulty. Again, with suitable design, these objects can be made to self-assemble to form larger and more complex objects. Problems on this path seem smaller in total, but are more interdisciplinary, requiring innovations both in synthesis and design.
3. Solution-phase mechanosynthesis guided by AFM mechanisms appears feasible, and could be used to build molecular objects containing $\sim 10^5$ highly crosslinked monomers without need for Brownian assembly. This path promises the greatest speed and flexibility of molecular construction, but its implementation requires still broader interdisciplinary collaboration.

Opportunities for hybrid approaches are evident. Biotechnology (and the stockpile of molecular components found in nature) can provide devices that can be incorporated into other structures. Solution-phase synthesis of medium-sized building blocks could provide components useful in mechanosynthesis. Molecular-tip AFM mechanisms could be used to characterize the products of biotechnological or chemical experiments, reducing the time required to detect and correct errors.

Each of the three approaches surveyed in this chapter grows out of experimental research already in progress; each can yield short-term scientific rewards while developing capabilities useful in building solution-phase molecular machine systems. Making new systems work in the laboratory is far more difficult than merely surveying possible approaches. By the same token, however, it is important to identify areas where successful experimental work can yield capabilities that open new fields of endeavor. If work is difficult, then it is best applied where the rewards seem large.

Some open problems

Much of this chapter can be taken as a description of open experimental problems. On the theoretical side, it would be useful to identify sets of compatible building-block molecules for use in solution-phase synthesis or in mechanosynthesis. This is a system design problem with numerous chemical constraints, not a problem of organic synthesis per se; it will not be solved by the traditional research strategy of synthesizing individual molecules having interesting structures or useful interactions with living organisms. Molecular CAD systems for the design of structures and syntheses based on such building blocks would be of great value.

5

MOLECULAR SORTING

Nanomechanical mechanisms and processes have natural applications to molecular manufacturing. As discussed in previous chapters that nanometer-scale structural components can have diverse shapes and good stiffness; that nanometer-scale gears, bearings chain drives, cams, and the like are feasible; that mechanical and electromechanical components can be combined to build systems of power-driven machinery capable of executing complex, programmed patterns of motion; and that mechanical manipulation of reactive moieties can be used to direct a broad range of synthetic operations with reliable positional control. The present chapter considers how these and related capabilities can be combined to acquire, order, and process feedstock molecules, first preparing reagent moieties and then applying them to workpieces to build up complex structures.

This process begins with small molecules in a liquid and ends with complex structures in an evacuated region. Liquid solutions are a natural starting point for molecular manufacturing: they are dense, thoroughly disordered, and can carry a wide variety of molecules. A vacuum environment is a natural destination: the use of vacuum avoids both viscous drag and mechanical interference from stray molecules, while minimizing unwanted chemical reactions. These characteristics improve the performance of molecular manufacturing systems and simplify their design and analysis. (For many applications, of course, the manufacturing system must ultimately deliver its product to a nonvacuum external environment.) To accomplish these steps, the required processes include:

1. Acquiring and ordering molecules from solution.

2. Transforming streams of molecules into streams of reagent moieties.
3. Assembling reagent moieties to form complex molecular objects.

Analyzing these processes requires attention to:

- Assembly processes
- Materials transport
- Cycle times
- Energy requirements
- Error rates
- Error sensitivities

SORTING

Nearly deterministic molecular manufacturing processes require well-ordered inputs, derived ultimately from a disordered feedstock. In designing a system, the contents of this feedstock can, within broad limits, be chosen for convenience. For concreteness, this section considers liquid feedstock solutions that can be prepared today: moderately pure mixtures consisting of small, relatively stable molecules of selected kinds in chosen proportions. (A typical small molecule, for present purposes, is one containing 20 or fewer atoms each of which has a degree of surface exposure equal to or greater than that of the central carbon atom in propane.) An input port to a manufacturing system can be presented with a feedstock consisting predominantly of a single chemical species (with trace contaminants) or with a mixture of species each present in substantial concentrations (again with trace contaminants).

The major species in a feedstock mixture can be chosen to serve as convenient fuels or raw materials; the trace contaminants must be assumed to be diverse. It is this diversity that poses the major challenge in analysis, forcing consideration of a set of molecules with a wide range of properties, and (in the early stages of processing) too large to be exhaustively described.

In constructing nanomechanical systems, only relatively exotic devices require isotopic refinement (e.g., for the construction of small, high-speed rotors with a precisely balanced mass distribution). Neither devices of this kind nor isotopic refinement are considered here.

This section first describes a class of mechanisms capable of selectively binding and transporting chemical species from a feedstock solution. It then discusses the combination of such mechanisms to build purification systems able to deliver streams of molecules that are (with high reliability) of a single type. Finally, it discusses the conditions that must be met in order to capture and orient single molecules from such a stream in a reliable manner, providing a eutactic input stream for subsequent processing.

Selective Transport

Basic concepts

Molecular biology provides numerous examples of systems that selectively bind molecules from solution (e.g., antibodies, enzymes, and various signal-molecule °receptors) and of systems that transport bound molecules against concentration gradients (e.g., active-transport systems in cell membranes). Nanomechanical systems can perform similar operations. The analysis of purification operations is fundamentally different from the analysis of processes in a eutactic environment, in that the population of impurity molecules is not specified. Accordingly, portions of this analysis are cast in terms of general properties of molecular shapes, although practical systems are likely to gain great advantages from the exploitation of other properties of particular molecules of interest.

There is a class of mechanisms capable of using mechanical power to transport molecules of a particular type from regions of lower to higher free energy, and from regions of lower to higher purity (see caption for description). Analogous mechanisms include devices with differing geometries (e.g., cam surfaces to the side of the rotor, rather than in the hub) and differing mechanisms for modulating the binding interaction (e.g., elastic deformation of the receptor, alteration of local charge distributions). The general capabilities of such devices can be analyzed using a combination of principles from statistical mechanics and experimental observations from molecular biology.

The capabilities of general structures can equal or exceed those of antibodies, about which much is known from experimentation. Antibodies can be developed that will bind anyone of a wide range of small molecules with small dissociation constants [typical values of

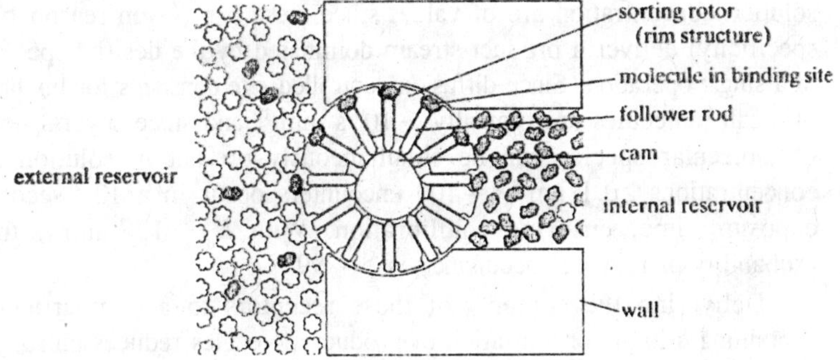

Fig. 5.1. A sorting rotor based on modulated receptors.

K_d are 10^{-5} to 10^{-11} nm^{-3}] and strong specificity (relatively poor binding of all but very similar molecules). The probability that a receptor will be found occupied by a molecule of the desired type is related to the concentration of a bind able molecule (ligand), C_{ligand}, and K_d by the expression

$$P_{occupied} = \frac{C_{ligand}}{K_d} P_{unoccupied} \quad ...(1)$$

Values of K_d depend on the interaction energies of the ligand and receptor, the ligand and solvent, and the solvent and receptor. Among differing solvent systems in which the interactions of the solvent with the ligands and the receptor are relatively weak and nonspecific, K_d will be relatively solvent independent.

The number of atoms required to equal the basic structural diversity of antibodies was estimated on combinatoric grounds to be <35. On geometric grounds, however, a receptor that encloses a molecule of substantial size (e.g., ~0.9 nm in diameter) with a layer of equal thickness must have an excluded-volume diameter of ~2.7 nm, and contain ~10^3 atoms. (The diameter of an active, Fv fragment of an antibody provides a highly conservative upper bound of ~4 nm on receptor size.) Allowing 2.7 nm of circumference per receptor, and 12 receptors per rotor yields a rotor diameter of ~10 nm. With a rotor thickness of 2.7 nm, the total number of atoms in the rotor system is ~2×10^4.

Modulated receptor mechanisms can be applied either to concentrate a molecular species from a relatively dilute feedstock solution, or to purify a stream of molecules by excluding impurities. For use in concentration, receptors having K_d considerably less than the actual solution concentration are of value, since these can (given reasonable specificity) deliver a product stream dominated by the desired species in a single operation. Since diffusion-controlled rate constants for binding of small molecules are typically ~10^9 s^{-1} nm^{-3}, and since several tens of molecular species can be simultaneously present in solution at concentrations ≥0.1 nm^{-3}, ~100 encounters occur in a 10^{-6} second exposure time, ensuring equilibration. With $K_d \leq 10^{-3}$ nm^{-3}, the probability of receptor occupancy is ≥~.99.

Delivering the contents of these receptors to a compartment containing a high concentration of product molecules reduces entropy, requiring a work of concentration amounting to ~20 maJ per molecule at 300 K. This can readily be provided by coupling to a mechanical drive mechanism. With the application of greater drive forces, a modulated receptor mechanism can easily pump against pressures of several gigapascal (i.e., several nanonewtons per square nanometer), at an energy cost on the order of 100 maJ per molecule. Where concentrations and pressures differ little from one side to the other, as in the later stages of purification cascades, the input work required can be quite small.

Diamondoid vs. protein structures for receptors

Diamondoid structures have advantages over proteins in their feasible affinities and specificities. In general, stiffer structures permit greater specificity. At one extreme, a sufficiently soft receptor can exhibit solventlike specificity, that is, very little. (Receptors containing flexible, oligomeric chains can provide mechanically modulated, broad-spectrum affinity.) Proteins can exhibit substantial stiffness, enabling strong specificity. Diamondoid structures can exhibit stiffness greater by several orders of magnitude, permitting significant improvements in specificity, particularly where exclusion of a particular species depends on overlap forces resulting from a bad geometric fit.

Diamondoid structures have a greater atom number density, leading to stronger van der Waals attractions and hence to greater binding energies, which can contribute to greater affinities. Further, the greater design-stage flexibility of general structures will often permit better matching of ligand and receptor, with respect to both geometry and electrostatics.

Specificity

Receptor affinity (that is, $1/K_d$) and specificity (roughly, differences in affinity for different molecular species) are related but distinct properties. In a practical system, what matters is the ability to discriminate between the desired molecule and the competitors *actually present* at that location in the process, not specificity for the desired molecule with respect to all possible competitors. Accordingly, the optimal receptor structure may differ at different stages in a cascade.

A high binding affinity for the preferred ligand can be a poor design choice if it leads to slow dissociation kinetics for a competitor, even if the differential in affinities remains large. Under these circumstances, the occupancy of receptors in transit through the barrier is controlled by competitive binding kinetics, rather than by equilibria; this typically results in poor discrimination. This problem does not arise in systems with $K_d > \sim 10^{-3}$ nm^{-3} and having $\sim 10^{-6}$ s receptor exposure times.

If two similar molecules can compete for the same receptor, then (neglecting differences in solvation) the ratios of their dissociation constants are directly related to the differences in their free energies of binding in vacuum and (neglecting pV effects and differences in entropy) are approximately related to the differences in their potential energies of binding in vacuum

$$K_{d,1}/K_{d,2} \approx \exp[(\Delta F_1 - \Delta F_2)/kT] \approx \exp[(\Delta V_1 - \Delta V_2)/kT] \quad ...(2)$$

The potential energy differences can readily be estimated by energy minimization in a molecular mechanics model; evaluation of the free energy differences requires calculations that either explore a region of the PES or perform an integration over a region.

Consider a receptor that, entirely surrounds its ligand (to permit the entrance and exit of ligands, the receptor must have at least one moving part). Assume that the receptor conforms closely to the contours of the preferred ligand, with the interfacial atoms placed near their minimum-energy distances, or somewhat closer. Alternative ligands can now be classified into three categories based on their size and shape: (1) those that resemble the preferred ligand, but occupy a strictly smaller volume in some conformation (e.g., having a similar structure, but with H substituted for CH_3); (2) those that resemble the preferred ligand, but occupy a strictly larger volume in all conformations (e.g., by having the reverse substitution); and (3) those of all other shapes. Ligands in category 1 typically bind more weakly, owing to loss of van der Waals attraction energy; those in category 2 bind more weakly (or are effectively excluded) owing to overlap repulsion; those in category 3 suffer both difficulties simultaneously, since they can neither fill the receptor nor avoid overlap.

Discrimination against ligands in category 1 increases with the Hamaker constant of the receptor structure; this constant is larger for typical diamondoid structures than for proteins. Even for proteins, however, the discrimination against ligands that lack a CH_3 group (putting H in its place) can be ~24 maJ, altering binding affinity by a factor of ~0.003 at 300 K.

Discrimination against ligands in categories 2 and 3 can be maximized by using a stiff receptor structure that surrounds the preferred ligand at less than the minimum-energy distance. The use of such *tight-receptor structures* sacrifices °binding energy in exchange for a particular kind of specificity, and may *reduce* discrimination against ligands in category 1. One of the smaller increments in steric bulk occurs between structures in which CH replaces N, shifting regions of similar overlap energy outward by ~0.09 nm over a small region. If the initial separation between an N atom in a ligand and an atom of the receptor is chosen to be ~0.3 nm, computational experiments in the MM2 approximation indicate that the adverse overlap energy experienced by the competitor can be ~45 maJ, sufficient to alter affinities by a factor ~5×10^4 at 300 K. (Note that in the present application, preferred ligand molecules can be selected with ease of purification as a criterion, hence structures presenting unusual difficulties can be avoided during the design stage.)

This example—discriminating between two species differing by the substitution of CH for N—also illustrates the utility of nongeometric mechanisms for molecular discrimination. Only small differences in shape arise from replacing a lone pair with a bond to hydrogen, but the resulting structures differ greatly in their electronic properties. In this instance, the nitrogen lone pair (but not the CH) can participate in the formation of a dipolar bond or a hydrogen bond. Geometrically, such bonds permit a closer contact between the N and the interface of the receptor. Energetically, they permit stronger binding. In general, different species with similar shapes have dissimilar electronic properties, permitting nongeometric mechanisms for discrimination.

Tight-receptor mechanism

A tight-receptor mechanism can be implemented. In this paired-rotor approach, all competing molecular species are in fast equilibrium with the fully exposed receptors on the left, but only the preferred species (and strictly smaller ligands) can pass through the fully enclosed position between the two rotors without encountering a large barrier caused by overlap repulsion. According to the estimate made barriers >30 maJ should be easily implemented. Tuning of the rotor potential energy function can effectively eliminate barriers when passing the preferred ligand (the common case for a purification process).

The paired-rotor, tight-receptor mechanism requires a compliant drive mechanism to operate properly. With an excessively stiff drive mechanism, even bulky competing ligands would routinely be forced through the gap (which must permit their passage in order to avoid failure by mechanical obstruction, regardless of drive stiffness). With a sufficiently compliant drive mechanism, however, the rotor moves (on short time scales) as if in free rotational diffusion. The rotational relaxation time for a rotor this size is $< 10^{-7}$ s, assuming a mean viscosity of the surrounding medium like that of water. Accordingly, in a system transferring receptors at mean intervals of 10^{-6} s, a bulky

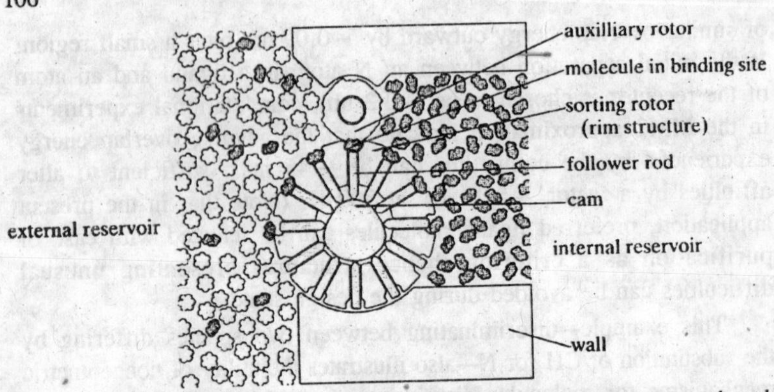

Fig. 5.2. A sorting rotor with an auxiliary rotor that forces bound molecules to pass through a totally enclosed state, effectively excluding molecular species that are unable to fit within a volume of defined size and shape.

competing ligand momentarily blocked by a barrier has ample time to be carried backward by several receptor diameters and escape into the surrounding solution. A nonlinear compliance can provide a small driving force (e.g., $\sim 10^{-12}$ N, measured at the rotor rim) with negligible stiffness over a certain angular range of motion (e.g., several radians), while providing a strong, stiff constraint that prevents larger excursions of the rotor relative to an underlying drive mechanism of ordinary stiffness. This appears to meet the relevant constraints.

Mechanisms of this sort can strongly discriminate against ligands unable to fit within the space occupied by the preferred ligand. This capability is exploited in the analysis of staged cascades for purification of input streams.

Energy dissipation

Properly designed sorting rotor systems can approach thermodynamic reversibility in the limit of slow motion when handling fluids of nearly pure composition (otherwise, entropies of mixture can be significant). Aside from speed-dependent (chiefly viscous) drag mechanisms, free energy is dissipated when potential wells in disequilibrium are merged. Well merging occurs in a system whenever a receptor is exposed to a molecular reservoir; to avoid energy dissipation during this process, the wells must, in each configuration as they are exposed, have a probability of occupancy that is precisely the probability of occupancy that they would have after a long, equilibrating period of exposure in that configuration. This condition is automatically achieved for well-blocked receptors in the outbound transportation process: their probability of occupancy is zero when exposed to the interior reservoir, and remains zero when they are exposed to the exterior reservoir. This condition can also be achieved for receptors during the inbound transportation process between two reservoirs containing known concentrations: a receptor with a particular configuration has a certain probability of occupancy as it leaves the external reservoir, and would (if unchanged) have a higher probability of occupancy on exposure to the internal, higher-concentration reservoir. A downward modulation of the affinity of the receptor during the inbound transportation process, however, can make these probabilities correspond. This can be achieved by adjusting the shape of the cam surface.

Viscous drag can be estimated from rotor dimensions, given some choice of effective viscosity for the surrounding solution environments. With the dimensions, the wetted area per rotor is $\sim 50\ \text{nm}^2$, and the characteristic length for shear (which is subject to considerable design-stage control) is ~ 2.7 nm. At a rim speed of 0.0027 m/s (10^6 receptor-sites per second), the drag power in a fluid of viscosity $\eta = 10^{-3}\ \text{N}\cdot\text{s/m}^2$ is $\sim 10^{-16}$ W, or ~ 0.1 maJ per receptor transferred. Sliding-interface drag in associated bearings can be made negligible in comparison.

The energy dissipation per operation can be reduced somewhat by using thick rotors with several rows of receptors, thereby increasing the characteristic scale length of fluid shear and reducing fluid structure effects. Dissipation can be reduced to a greater extent by increasing the number of rotors (or the number of rows of receptors per rotor) and slowing the rotor speeds in proportion. If a single-row rotor has $\sim 2 \times 10^4$ atoms, and a rotor with its housing and prorated share of the drive system contains $\sim 10^5$ atoms, then at 10^6 operations per second, it processes its own mass in < 0.1 s. A considerably lower frequency would remain compatible with high productivity.

Damage mechanisms and lifetime

Modulated receptors for concentrating and purifying molecules differ from other mechanisms considered in this volume in that they are, by the nature of their task, exposed to an ill-defined chemical environment. Their susceptibility to damage caused by chemical reactions varies with structure and environment, and can most easily be estimated by analogy to other ill-defined chemical systems.

The surface reactivities of diamondoid structures (if designed with low surface strain, etc.) resemble the reactivities of smaller organic molecules. The stability and lifetime of solvent molecules in relatively pure, unreactive solutions can thus provide a model for the stability and lifetime of small patches of surface structure, including receptors, exposed to a reasonably pure, unreactive feedstock solution. This suggests reasonably long operational lifetimes.

Damage resulting from trace quantities of highly reactive contaminants can be minimized by flowing feedstock solutions past surfaces bearing bound moieties resembling those used on critical rotor surfaces, but selected for higher reactivity. Sacrificial moieties of this sort can combine with and neutralize many reactive species, including free radicals.

Proteins in living systems provide a model for molecular machines in a relatively complex, chemically aggressive environment. Metabolic enzymes can have lifetimes of several days, despite the relative fragility of protein structures. Lifetimes of diamondoid sorting rotors of greater stability in a more benign environment should be longer, but even a one-day lifetime is sufficient for a device to process -10^6 times its own mass before requiring replacement.

Cascades of Modulated Receptors

To achieve low error rates in a molecular manufacturing process requires an input stream with a low fraction of contaminant molecules. For example, if a device is to perform 10^6 operations per second for several decades without an error, the fraction of contaminant molecules must be held to $\leq 10^{-15}$. This degree of purity can readily be achieved (for a wide range of reasonably stable feedstock molecules) by using a cascade of modulated receptors.

Multi-stage cascade systems are ubiquitous in chemical engineering practice, with implementations including leaching systems, distillation columns, extraction columns, and the like. The analysis of cascades is in general complex. The following section presents an analysis of a simple case that illustrates the essential properties of cascades for the present application.

A model of staged cascade systems

Cascades combine multiple separation stages in a counterflow system. For simplicity, consider a system with N reservoirs of fluid containing predominantly molecules of a single type (the desired product), but contaminated by small concentrations $c_0,\ c_1,\ c_2, \ldots C_{N-1}$ of impurity molecules (measuring concentrations in molecule fractions). Each reservoir n (save the last) is linked to reservoir $n + 1$ by an inbound flow (from the source to the destination) carrying f_{in} molecules

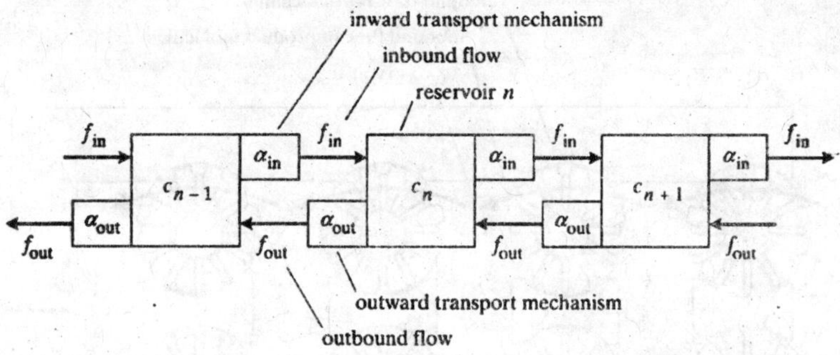

Fig. 5.3. Schematic diagram of a portion of a staged-cascade process.

per second, and each save the first is linked to reservoir $n - 1$ by an outbound flow carrying f_{out} molecules per second. (This is a simple instance of a flow pattern that permits mass balance in each reservoir.) Each flow is the product of a selective transport mechanism characterized by a differential affinity for impurity and product molecules; the inbound stream from compartment n thus has an impurity concentration $c_n\ \alpha_{in}$, and the outbound stream has an impurity concentration $c_n\ \alpha_{out}$. The condition for mass balance for impurities in reservoir n is then

$$c_{n-1}\alpha_{in} + c_{n+1}\alpha_{out}f_{out} - \left(c_n\alpha_{in}f_{in} + c_n\alpha_{out}f_{out}\right) = 0 \qquad ...(3)$$

The assumption of stage-independent values for f_{in}, f_{out}, $\alpha_{in,}$ and $\alpha_{out,}$ together with a constant value of c_0, leads to a steady-state solution in which c_n declines exponentially with n. Convenient variables are then

$$f_{out} / f_{in} = f_{rat};\ \alpha_{out} / \alpha_{in} = \alpha_{rat};\ c_n / c_{n-1} = c_{n+1} / c_n = R \qquad ...(4)$$

where the value of R, which defines the rate of the exponential decline, is determined by

$$R = \left(f_{rat}\alpha_{rat}\right)^{-1} \qquad ...(5)$$

The final impurity concentration is then

$$c_N = c_0 R^N \qquad ...(6)$$

which for R « 1 can be extremely small for modest values of N. Although real systems will never correspond precisely to this model, it can be used to estimate the relationships among receptor properties, system size, relative flow rates, and product purity.

Purification using staged cascades of modulated receptors

For concreteness, consider a system that must deliver a stream of product molecules with a contaminant fraction $\leq 10^{-15}$, starting with a

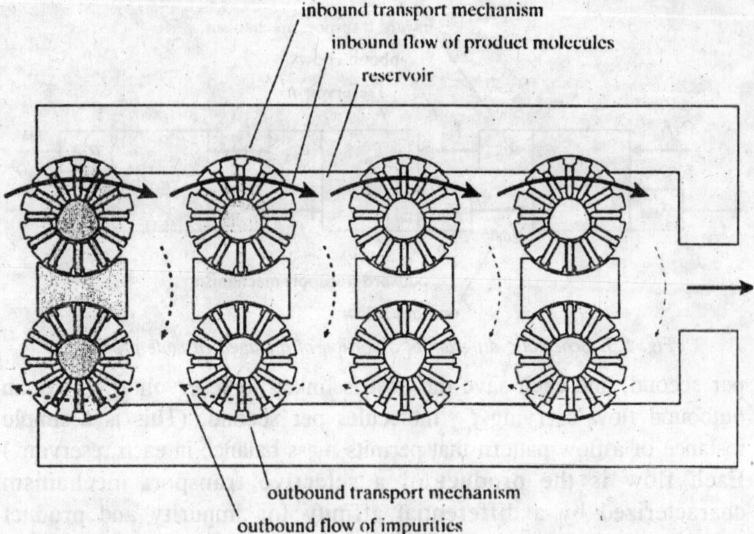

Fig. 5.4. Schematic diagram of a staged-cascade process based on sorting rotor.

feedstock mixture containing product molecules at a fractional concentration of 10^{-2} ($\sim$0.1 nm^{-3}). For generality, assume that only the universally available geometric properties of molecules are exploited to achieve receptor specificity, neglecting the use of electrostatics, hydrogen bonds, and so forth (all of which play prominent roles in binding by proteins). Although the staged-cascade model assumes that the selectivity factor α_{rat} is a constant, the presence of a mixture of impurities complicates this picture, motivating the use of different receptor structures and specificities at different stages.

Receptors for the inward transport mechanism of the first stage are best designed for high product affinity in the presence of a dominant background of quite different molecules. Analogies with antibodies suggest that an inward-transport rotor can deliver a product stream with impurity fractions of $\sim 10^{-4}$ to 10^{-9}, depending on affinities, specificities, and the concentrations of the effectively competing ligands. The value of α_{rat} for the transport stage depends on the specificity of the outbound mechanism. If this is completely unselective, and if f_{rat} = 0.1 (corresponding to a modest outbound flow rate), then the value of R for this stage can range from $\sim 10^{-3}$ to 10^{-8}, yielding impurity fractions of similar numerical magnitudes.

Receptors for the subsequent purification stages should usually be designed with lower product affinities, to ensure equilibrium rather than kinetic control of transport selectivity. With tight-receptor mechanisms, the concentrations of competing ligands having excess bulk is reduced rapidly, with feasible values of $R \geq \sim 5 \times 10^3$, given $f_{rat} = 0.1$. Ligands substantially smaller than the product molecule are effectively discriminated against by loss of van der Waals attraction energy. After several stages of this sort, the chief impurities will typically be molecules that resemble the product, save for lower steric bulk in some region. The number of molecular types that can stand in this relationship to an already-small product molecule are few. Low values of R can be ensured in this circumstance by use of several selective, tight-receptor mechanisms in the outbound direction, each adapted to bind a specific impurity species. With transport mechanisms adapted to each stage as just discussed, values of $\alpha_{rat} \geq 10^4$ should be readily achievable throughout, hence analyses using $R = 10^{-3}$ will yield conservative results. Accordingly, producing a stream with an impurity fraction $\leq 10^{-15}$, starting with a feedstock with an impurity fraction $\sim .99$, should require a number of stages $N \leq 5$.

Overall performance

To construct a system with relatively large throughput, many sorting rotors can be used in parallel, each with multiple rows of modulated receptors; the number of rotors in a set can then be adapted to the magnitude of the stream of molecules to be handled, and operating rates can be equalized. Parallel systems lend themselves to the construction of redundant architectures in which the useful system lifetime is comparable to the mean lifetime of the individual rotors. Rotors with multiple receptor rows can also reduce the viscous drag loss per receptor.

The rotors handling the inward flow streams are the dominant source of energy dissipation, assuming low values of f_{rat} as above. For $N = 5$ and capacities of $\sim 10^6$ molecules per second per row, losses from this mechanism amount to ~ 0.5 maJ per molecule delivered. This is small compared to the useful work performed in molecular concentration (~ 20 maJ per molecule); losses from entropy of mixing are greater than this, amounting to $\sim 10\%$ of the useful work (assuming $f_{rat} = 0.1$). With the estimated per-rotor masses (etc.) a purification system can deliver a quantity of product equaling its own mass in <0.5 s.

Handling harder cases

Separation on the basis of shape will fail (or, more properly, require many steps) if shapes are nearly identical. Among small molecules that consist of the elements of greatest interest, differences in atomic valence, bond length, and nonbonded radius guarantee substantial differences in shape. Among molecules containing transition or lanthanide metals, however, these differences can become quite subtle. In these instances, alternative strategies relying more heavily on chemical and electronic properties (e.g., differential complexation of ionic species, as in chromatography) may be desirable; laboratory experience demonstrates that these elements can be separated in this way. In manufacturing systems consuming the mix of elements expected to be typical, separation of these more difficult elements will be a quantitatively small portion of the overall processing task, having little impact on such system parameters as mass, energy consumption, and productivity.

Ordered Input Streams

Sorting processes of the sort can keep a reservoir filled with molecules that are reliably of a single type. To provide deterministic inputs to a molecular manufacturing process, these molecules must be transformed into a stream of bound moieties such that each site in the stream is reliability occupied. This could be accomplished by combining an unreliable mechanism for ordering and reactively binding molecules with a mechanism for probing and sorting the resulting structures; this section will instead describe approaches capable of achieving reliability directly.

Steps in ordering and input

One approach to providing ordered inputs comprises the following steps:

1. Expose a receptor to a pure liquid under conditions that ensure reliable occupancy of that site by a bound molecule.
2. Expose the bound molecule to a reagent moiety (on a separate transport mechanism) under conditions that ensure the reliable formation of a specific covalent bond.
3. Transfer the molecule to the other transport system as a covalently bound moiety, emptying the receptor for reuse.

This approach to molecular ordering with transfer to a reagent belt mechanism. In a mechanism of this sort, step 3 is a natural consequence of the pattern of motion, leaving steps 1 and 2 as the primary design problems.

Ensuring reliable receptor

A receptor will be occupied reliably ($P_{empty} \leq 10^{-15}$) if the free energy of binding has a magnitude greater than ~143 maJ. The loss

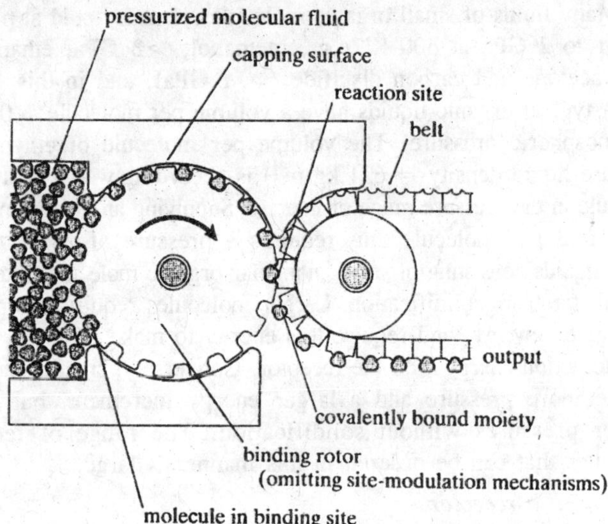

Fig. 5.5. Schematic diagram of a mechanism for removing molecules from a liquid phase and covalently binding them to a moving belt.

of entropy that occurs when a molecule is bound from solution is comparable to the loss of entropy in freezing to a crystal; for small organic molecules, a typical value is 6×10^{-23} J/K, hence the required potential energy of binding is $\Delta V_{binding} \approx -161$ maJ.s

Two sources of driving force can be applied to increase $\Delta V_{binding}$: attractive intermolecular forces to reduce the energy of the bound molecule, and hydrostatic pressure to increase the Gibbs free energy of the unbound molecules. In general, feasible binding energies increase in magnitude with increasing molecular size, polarizability, and dipole moments. Feasible hydrostatic pressures (i.e., those that do not cause the liquid to solidify) tend to decrease with the same variables. The magnitudes of feasible binding energies can be evaluated using molecular mechanics methods; the magnitudes of feasible hydrostatic pressures are indicated by data on liquids at high pressures.

A model receptor for a small nonpolar molecule, ethyne; molecular mechanics calculations indicate $\Delta V_{binding} = -97$ maJ (this example is chosen to illustrate receptor principles with a simple, symmetrical structure, not necessarily to recommend ethyne as a feedstock). A larger model (including more atomic layers) would increase van der Waals attractions, increasing the magnitude of the binding energy. Achieving $\Delta G_{binding} = -161$ maJ requires an increment of 64 maJ; this can be supplied by hydrostatic pressure.

Many fluids of small organic molecules remain liquid at pressures of ~1 to 2 GPa at 300 K (e.g., methanol, >3 GPa; ethanol, >2 GPa; acetone and carbon disulfide, > 1 GPa), and in this pressure range, typical organic liquids have a volume per molecule ~0.75 that at atmospheric pressure. The volume per molecule of ethyne (low-pressure liquid density = 621 kg/m^3) is accordingly ~5 × 10^{-29} m^3/molecule in the pressure range of interest. Supplying an energy increment of 64 maJ per molecule thus requires a pressure of ~1.3 GPa, at which liquids consisting of small, nonpolar organic molecules like ethyne are still far from solidification. Larger molecules require less pressure for a given energy, and require less energy to make up for deficits in the interaction energy with the receptor. Conversely, smaller molecules require more pressure and a larger energy increment, but tolerate greater pressures without solidification. The range of feedstock molecules that can be ordered in this manner is large.

Bonding in a receptor

For suitably chosen feedstock molecules and receptor designs, it can be assumed that the molecule is either well oriented, or that patterns of symmetry and reactivity make molecular orientation unimportant. The properties of the receptor itself need not constrain the reaction environment: the bound molecule is confined by moving under the capping surface, and can then be subjected to direct, forcible mechanical manipulation. Such processes can move it to a new location characterized by high free energy and stiff intermolecular contacts.

The reaction that transfers the molecule must take the form of an addition (unless means are provided for handling displaced groups). For unsaturated molecules, radical additions or Diels-Alder cycloadditions are candidates. For saturated molecules, additions to transition metals are candidates. To overcome activation energy barriers, large piezochemical forces can be applied. Strategies for forming strong bonds with little energy dissipation are available. The principles underlying these processes are the same as those in the mechano-synthetic processes. In working with small molecules, however, it is difficult to provide stiff restoring forces that resist motion of the molecule in the direction of the attacking moiety. This limitation may, for example, preclude low energy dissipation in radical coupling reactions involving small molecules.

As is true throughout the design of materials input subsystems, any candidate feedstock moleculc that presents an intractable problem in some regard can be discarded in favour of an alternative. Feedstock molecules can be chosen to provide the necessary elemental inputs in convenient forms.

Molecular Mills

It is assumed that accurately controlled mechanical motions can be imposed as a boundary condition in molecular systems, implying control of positions, trajectories, and forces during chemical transformations. This section describes mechanisms that can provide such control using systems of belts and rollers; It will further describe more flexible mechanisms based on programmable manipulators.

Reactive Encounters

Systems of belts and rollers can be used to implement mill-style molecular processing systems. Belts moving over rollers and along guides find widespread use in macroscale manufacturing: they transmit power and transport materials and parts. They can serve this function in nanoscale systems as well. With the help of auxiliary mechanisms—such as moving parts in reagent devices, rollers and cam surfaces to drive those parts, devices for transfer to and from other transport systems—molecular mills built in this fashion can perform a wide range of mechanochemical operations within a larger manufacturing system.

Simple encounters

The simplest reactive encounters involve direct contact and separation, a motion which can be provided by a device. An encounter of this sort can transfer a group through an abstraction reaction or through a more complex chemical mechanism. Among the mechanical parameters subject to design control are the peak compressive load (which can be $>$ 10 nN), the stiffness of reagent moieties relative to one another along the line of motion ($\sim$ 90 *N/m* with sufficiently stiff bearings), and the relative stiffness of the moietal supporting structures perpendicular to this line (with pairs of reagent devices having sterically complementary interfaces placed under compressive load, this can easily exceed *50* N/m). An encounter of this sort can be performed with a belt-belt contact, a belt-roller contact, or a roller-roller contact.

The reaction time depends on belt speed, moietal trajectories (determined by the belt and wheel geometries), and the range of positions regarded as comprising the encounter state. Assuming that the trajectories are characterized by simple circular motions of radii r_1 and r_2 at a speed v with a positional tolerance of $\pm\delta$ (assumed small), the reaction time is

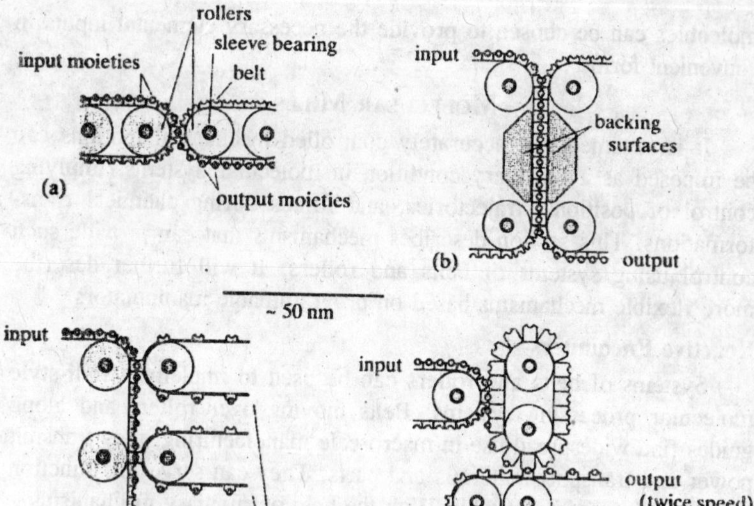

Fig. 5.6. Schematic diagrams of reactive encounter mechanisms, providing for simple contact and separtion (a), prolonged contact with opportunity for substantial cam-driven manipulation (b), and transformations of moiety transit frequency (c) and speed (d).

$$t_{\text{trans}} \approx \frac{2}{v}\left(r_1^{-1} + r_2^{-1}\right)^{-1} \cos^{-1}\left(1 - 2\delta\left(r_1^{-1} + r_2^{-1}\right)\right) \approx \frac{4}{v}\sqrt{\delta\left(r_1^{-1} + r_2^{-1}\right)} \quad ...(7)$$

For $v = 0.005$ m/s, $r_1 = r_2 = 10$ nm, and $\delta = 0.01$ nm, ($t_{\text{trans}} \approx 1.8 \times 10^{-7}$ s, moderately greater than the typical value. Fast, simple-encounter reactions could be performed with higher speeds and smaller mechanisms. Reaction times can be indefinitely prolonged by placing belts in contact and passing them between a pair of backing surfaces.

Complex encounters

In complex reactions, a device must drive several coordinated, properly sequenced motions. This can be accomplished by elaborating the structure of the reagent device to include moving parts actuated by auxiliary rollers or cam surfaces, while extending the encounter time (if need be).

This approach enables thorough control of molecular trajectories and reaction environments. As the two reagent devices come together, they can entirely surround the pair of moieties with a eutactic environment. Multinanometer-scale reagent devices containing multiple moving parts can manipulate this environment, forcing the moieties and adjacent structures to execute motions driven by stiff mechanisms capable of exerting large forces. Cam surfaces adjacent to the belts can interact with one or more protruding cam followers on the device to guide and power an arbitrarily long sequence of arbitrarily complex motions. Relatively small and simple mechanisms in this class are capable of exploiting pi-bond torsion effects, performing three-moiety encounters such as regeneration of alkynyl radicals assisted by radical displacement, and so forth.

A complex encounter mechanism can also exploit non bonded interactions to force the transfer of a group. For example, a bulky structure joined to another by a single bond can be mechanically clamped and then pulled to force cleavage of the bond and creation of a pair of radicals.

Conditionally repeated encounters

As discussed earlier the use of conditional repetition to convert a reaction that on the basis of single-encounter thermodynamics and kinetics has a probability of failure P into a reaction process with a smaller probability of failure p^n, where n is the potential number of encounters. Conditional repetition can both increase reliability and (by permitting the use of less exoergic reactions) decrease energy dissipation.

The specific geometry and kinematics represent only one embodiment of a broad class of analogous mechanisms. For successful

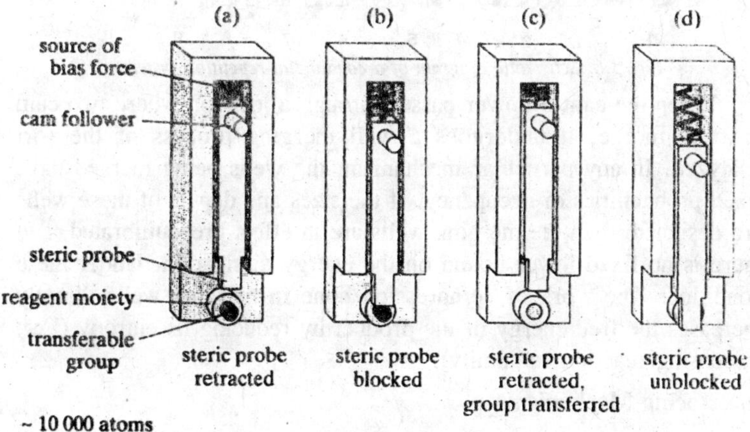

Fig. 5.7. Schematic diagram of a reagent device for implementing a conditional-repetition strategy to increase reaction reliability.

operation, the probability of the probe moving to its extended position [a state like (d)] must be negligible (e.g., $\leq 10^{-15}$) while the transferable group is still present. The driving force for this motion can be small (e.g., ~0.1 nN), and the formation of a bond between the probe and the site exposed by transfer of the group is among the acceptable outcomes, so long as that bond can be broken by the forcible retraction of the probe.

The net effect of which is to attempt the transfer reaction if and only if the group has not yet been transferred. As illustrated, the blockage of further attempts after a successful transfer can be direct, by blocking the reactive site; the motion of the probe structure could equally well achieve the same result though indirect mechanical linkages. Note that analogous mechanisms can be devised that make reaction attempts contingent on the *absence* of a group, hence the probe assembly need not be on the source side of a reactive encounter system.

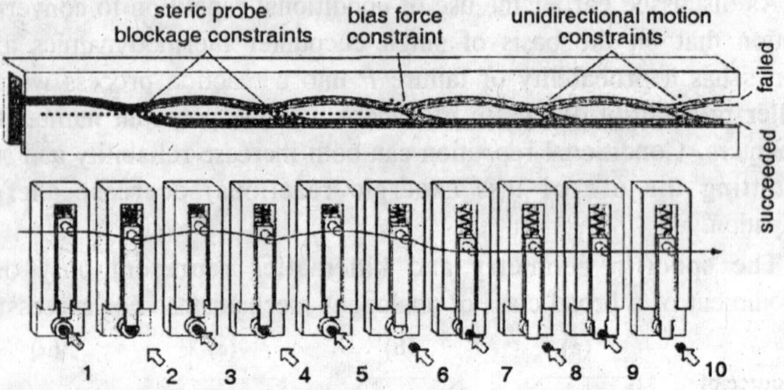

Fig. 5.8. Schematic diagram of a conditional-repetition system.

When the cam follower passes through a location where two cam grooves merge, it undergoes a well merging process of the sort analyzed. In any particular mechanism, the wells being merged have fixed probabilities of occupancy. If the sizes and depths of these wells are designed, then the merging wells are in effect preequilibrated, and there is no fixed lower bound on the energy dissipation. Under these conditions, the work of compression done in merging wells directly increases the free energy of the product by reducing its entropy (i.e., increasing reaction reliability).

Interfacing Mechanisms

Frequency and speed

System-level flexibility requires several kinds of transformations among streams of reagent moieties in molecular mill systems. Where several consumer subsystems require inputs of moieties at *low* frequencies, but a producer subsystem can efficiently produce them at high frequencies (or vice versa), it is desirable to transform frequencies by splitting and merging streams. Differing subsystems can have widely differing optimal speeds of belt motion.

Nondeterministic interfaces

In subsystems built using devices of the sorts described thus far, the rate of motion and frequency of operation at any given point fully determine the rates and frequencies at every other point (save for transient fluctuations). In building larger systems, it is desirable to surround subsystems of this sort with interfaces permitting greater flexibility of operation. In particular, this is required for fault-tolerant operation, which in turn is required in building relatively large and durable systems.

Transfer of a stream of moieties or product structures to a stream of pallets, moving along a track but not linked to form a belt, can provide the necessary flexibility. In regions where pallets are close packed, a stream of pallets can be treated like a moving belt, and similar transfer operations can be performed. In other regions, pallets can be more or less closely spaced (permitting the accumulation and use of buffer stocks), and the tracks bearing them can split or merge.

Two classes of track junction are useful in the design of fault-tolerant systems. These are fair, nonblocking merging junctions, and fair, nonblocking distribution junctions. A merging junction (1) accepts at least two input streams of pallets and produces a single stream of output pallets that includes contributions from both inputs (i.e., it is fair), and (2) continues to produce a stream of output pallets so long as one input stream has not failed (i.e., stopping inputs does not block the node). Similarly, a distribution junction (1) accepts a single input stream and produces one or more output streams, distributing the inputs across all the outputs, and (2) continues to distribute inputs to an output so long as one output stream remains unblocked. A *failed* junction of either kind is assumed to block one or more of its inputs; in the worst case, failure blocks all inputs (and thus outputs) by blocking all motion through the junction. These simple abstract behaviours can be implemented with active mechanisms, and perhaps without them.

A network of mechanosynthetic devices linked in this manner can behave as a demand-driven supply system. The combination of applied drive forces and mechanochemically derived forces from exoergic reactions can bias devices toward outputting product, but their output can be blocked when space for more product becomes unavailable. Production is then paced by the rate of removal of products at the end of the network, and no computational synchronization of production with demand is necessary. This somewhat resembles just-in-time delivery systems in conventional manufacturing, in which empty containers are the signal to produce more parts, but it more nearly resembles the delivery of water by a pipe, in which removal of material at one end creates a pressure gradient that moves material throughout the length.

Reagent Preparation

Molecular mills can be used to transform low-entropy streams of feedstock molecules into low-entropy streams of reagent moieties covalently bound to reagent devices. Each of the group-transfer reactions in a process of this sort can occur in an enclosing environment designed to facilitate that reaction, to minimize misreactions, and to minimize energy dissipation.

A typical input molecule might be ethyne; a typical reagent moiety might be a strained alkyne. The intermediate stages in this process might include two radical additions and hydrogen abstractions, followed by the regeneration of the abstraction tools in a process that ends with the elimination of molecular hydrogen from a transition metal. Each of these steps (save the last) requires an encounter between two reactive groups, and each results in the transfer of some number of atoms between one site and the other. This example suggests that cycles that prepare reagent devices by charging them with fresh reagent moieties may typically require on the order of 10 reactive encounters.

Reagent Application

Molecular mills can be used to apply reagent devices to work pieces. Although programmable manipulators can do this with greater flexibility, producing many different products using one mechanism, mills can produce a single product with greater efficiency. Architectures for flexible manufacturing of macroscale products will likely include both mill-style and manipulator-style assembly mechanisms.

Reagent application closely parallels reagent preparation, except that the workpiece-side structure in each reactive encounter is subject to the constraint that it be an intermediate in the construction of a desired device. This precludes thorough adaptation of the reaction environment for the purpose of facilitating reliable, low-dissipation operations. Nonetheless, the freedom to choose designs that can in fact be manufactured and synthetic sequences that generate relatively favourable intermediate structures provides considerable latitude for avoiding unfavourable reaction environments.

Size and Mass Estimates

The masses of components in a mill-style processing mechanism are significant both in estimating system productivity and in estimating radiation-damage lifetimes. As suggested the structure (and hence the mass) of reactive encounter mechanisms can vary widely; to make a rough estimate of the typical system mass on a per-mechanism basis, a standard density can be assumed for all components (here, 2500 kg/m^3 or ~125 atoms/nm^3), and reference sizes can be allocated to provide for a moderately complex process and for transport of moieties over a moderately long distance between processes.

Encounter mechanism

Roller pairs and backing-surface pairs can have comparable masses; the roller option is considered here. A radius of 5 nm and a mean thickness of 2 nm provide ample room for a stiff bearing of the sort. A pair of rollers, including axles, has a volume of ~310 nm^3.

A moderately complex encounter process can be driven by a mechanism occupying a volume 4 nm on a side, or 64 nm^3 (four times the volume of the exemplar interlock mechanism). This mechanism can be formed by the mating of reagent devices from each of two facing belts, each of which can then be 4 × 4 × 2 nanometers in size. An auxiliary cam surface of substantial complexity can fit within a space 4 × 2 × 10 nanometers in size, adding 80 nm^3.

These estimates are appropriate for reagent preparation processes, where only small moieties are handled. They can serve reasonably well for reagent application processes, so long as the workpiece is limited to a diameter of several nanometers.

Transport mechanisms

A mean belt length of 20 roller-radii (= 100 nm) per roller provides for substantial flexibility of system layout. If each belt is a strip with a cross sectional area of 0.5 nm 2, then the volume is 50 nm^3 plus that of one or more reagent-device pairs. If each belt is a close-packed chain of 4 × 4 × 2 nanometer reagent devices, the volume of belt structure per roller-pair is 1600 nm^3. In the latter case, there are 50 reagent moieties per reactive encounter mechanism.

In a reasonably densely packed system, the distances between encounter mechanisms can be short; they can be stacked in parallel in near contact. A supporting structure with a mass equal to that of five struts of 4 nm^2 cross sectional area, each 15 nm (= 3 roller radii) in length seems ample; this amounts to 300 nm^3 of structural material per roller pair.

Totals

The total volume, mass, and atom-count per reagent device depend substantially on the choice of reagentdevice density along the belts. At the upper extreme, where the belt consists of a chain of close-packed devices, the numbers are ~2300 nm^3, ~5.7×10^{-21} kg, and ~2.9×10^5 atoms. Systems of this sort have relatively high throughput and energy efficiency, and will serve as the basis for later calculations unless otherwise indicated. At the lower extreme, with only one pair of reagent devices per roller pair, the numbers are ~800 nm^3, ~2.0×10^{-21} kg, and ~1.0×10^5 atoms. Systems of this sort have lower throughput and efficiency, but less mass per kind of operation performed (i.e., greater versatility per unit mass). For comparison, 2.0×10^{-21} kg is ~40 times the mass of many enzyme molecules.

These estimates are substantially larger than would be appropriate for a simple encounter mechanism and substantially smaller than would be appropriate for a conditionally repeated complex-encounter mechanism, but should not grossly underestimate the size and mass required per encounter mechanism in a mill-style processing system handling small reagent moieties. Assuming an arbitrarily chosen filled-volume fraction for the more massive system of ~10% yields a crude estimate of the total volume per reactive encounter mechanism, ~2×10^{-23} m^3.

Reagent processing system

Systems for transforming a stream of small feedstock molecules into a stream of small reagent moieties can be expected to require ~10 reactive encounters. Combining this estimate with those yields ~5.7×10^{-20} and ~2.0×10^{-20} kg, respectively, or ~2.9×10^6 and ~1.0×10^6 atoms. These estimates indicate that high-throughput reagent processing systems can produce their own mass in deliverable moieties in ~ 3 s. Since reagent application adds a single step to a sequence of ~10, estimates of the mass and productivity of a system producing delivered moieties are little changed.

Error Rates and Fail-stop Systems

Errors in molecular mills can arise from damage to the mechanisms of the mill, from the instability of reagent moieties, and from failed reactions (i.e., omitted reactions and misreactions). Conditions that suffice to ensure that errors resulting from misreactions, omitted reactions, and reagent instabilities occur at rates $\leq 10^{-15}$ per operation. The large stiffnesses feasible in mill mechanisms can help in meeting these conditions. Even with the exploitation of diamondoid, eutactic environments and stiff mechanochemical processes, however, these conditions exclude many potential reagents and transformations from use in reliable mechanosynthesis. It will here be assumed that choices compatible with these constraints are made, and that error rates in molecular mill systems are accordingly dominated by radiation damage. The following sections consider the consequences of damage, and describe how those consequences can, at the subsystem level, be made to result in a simple cessation of activity.

Consequences of damage

In a molecular mill, significant damage can be defined as damage that causes a failed reaction. In extreme instances, this may occur because damage has physically blocked the motion of the mill. In marginal instances, this may occur because damage has altered component geometries enough that misalignments increase the error rate.

In the former case, damage stops the mechanism directly; in all other cases, the mechanism continues to move for some time after the failed reaction. After a failed reaction, however, one or more of the departing product moieties will have an incorrect structure. An incorrect structure will usually cause failures in further reactive encounters, causing a cascade of damage transmitted from reagent devices to other reagent devices. Accordingly, significant damage will rapidly result in widespread damage.

A damage cascade could be prevented by measurement and correction systems following each reactive encounter mechanism, but the complexity of doing so in general appears excessive. Conditional-repetition mechanisms are formally members of this class, but they correct only a single, well-defined kind of error: an omitted reaction.

Error detection

After a reaction, a processing system can gauge the shape of each moiety and product using a measurement mechanism based on a roller with a shape complementary to that of the correct structure. If the actual structure has a protrusion where the correct structure would not, then the roller (if it has a small contact area) is typically forced outward by a distance comparable to the size of the protrusion. Note that failed reactions will usually be signaled by a protrusion on one of the output structures: the atoms that enter the reactive encounter must leave again, and a misplaced atom will typically occupy space that would otherwise be empty (if the space is internal, detection of the event must be indirect). A single gauge mechanism can test a reagent device or a substantial fraction of the area of a workpiece. Gauge mechanisms can reliably discriminate between correct and incorrect structures, using iterated probing if necessary. Detection of an error can trigger a mechanism that blocks further activity in a molecular mill subsystem.

The number of gauge mechanisms necessary to detect an error cascade in a large molecular mill subsystem is a small fraction of the number of reactive encounter devices, provided that an error can safely be allowed to remain undetected for many operation cycle-times. If products from each subsystem must transit a relatively long belt before being delivered to a consumer subsystem, then a relatively tardy detection process can prevent the delivery of mismanufactured product structures. The resulting *fail-stop* subsystems make convenient building blocks in the design of redundant, fault-tolerant manufacturing systems.

Acceptable subsystem masses

Redundant, fault-tolerant systems can be constructed from sets of subsystem units having identical functions, if these are linked in such a way that the essential input-output properties of each set are unchanged when one or more of its units fails. With cumulative damage (no repair is assumed here), system lifetime remains finite, limited by the lifetimes of the units.

Estimates of energy Dissipation

Energy dissipation in mechanochemical processes results from both mechanical and chemical steps. Mechanical energy dissipation results from the motions involved in transport and moietal manipulation; chemical energy dissipation results from the chemical transformations themselves.

Mechanical operations

Low energy dissipation is of greatest importance in systems designed for high throughput, and such systems will typically have belts with closely spaced reagent devices. Adopting a 4 nm spacing, a mechanism that delivers 10^6 moieties per second requires a 0.004 m/s belt speed; this speed is assumed throughout this section. Assuming a 5 nm roller radius, a 2 nm bearing radius and length, and 1000 N/m bearing stiffness yields an estimated power dissipation per roller of $\sim 7 \times$

10^{-20} W, using the more adverse parameters for band-stiffness scattering. With ~20 rollers in the system, this corresponds to ~1.4×10^{-24} J per delivered moiety. This is negligible in comparison both to chemical energies and to kT_{300}.

Dissipation is larger where belts pass between backing surfaces. With assumptions analogous to those made for the bearing interface, a 16 nm^2 reagent device sliding over a surface at 0.004 m/s with an interfacial stiffness of 1000 N/m dissipates ~10^{-19} W, or ~10^{-24} J per pair sliding a distance of 20 nm. This remains quite small.

The motion of cam followers through cam grooves causes phonon scattering by the mechanism. Scaling from the sample calculation in that section yields an estimated energy dissipation of ~10^{-27} J per nanometer traveled. Thermoelastic damping is of the same order, assuming forces of ~1 nN applied to areas of ~1 nm^2. Since these values are ~10^{-6} kT_{300} per nanometer of sliding motion, extensive use of cams is compatible with low energy dissipation per operation.

Reactive encounters can dissipate free energy through nonisothermal compression of potential wells. With good design practice, typical values at the speeds assumed should commonly be $\leq$0.1 maJ per cycle. Dissipation from this source can easily exceed that from all other mechanical mechanisms combined. A reasonable estimate of the total energy dissipated by mechanical operations in a ten-step mechanochemical process is therefore ~1 maJ per moiety delivered; if nonisothermal compression losses are minimized, the energy dissipated can be reduced.

Chemical transformations

A generous estimate of the energy dissipated in a series of chemical transformations can be made by assuming that each operation is as exoergic as the combustion of carbon, ~650 maJ per carbon atom. If each atom passes through ten such steps on the way to its final state, then the manufacture of a carbon-rich diamondoid product dissipates ~3×10^8 J/kg in the form of waste heat. A less generous but still naively high estimate can be made by instead assuming that each chemical transformation must be sufficiently exoergic to make it highly reliable ($P_{err} \leq 10^{-15}$); this would require the dissipation of ~145 maJ per step, yielding an estimate of ~7×10^7 J/kg. A more sophisticated estimate requires attention to ways in which good design practice can minimize energy dissipation. (Note that the naive estimate may in fact be reasonable for systems where low energy dissipation is not a design objective.)

Reliable mechanochemical operations can in some instances approach thermodynamical reversibility in the limit of slow motion. At 10^6 operations per second, many processes in this class dissipate energies in the range characteristic of mechanical motions. In comparison to 145 maJ, these energies are negligible. Accordingly, if one assumes the development of a set of mechanochemical processes capable of transforming feedstock molecules into complex product structures using only reliable, nearly reversible steps, then the energy dissipated in millstyle manufacturing processes could approach zero, and in practice be ~1 maJ per atom, or ~10^5 J/kg. (The conversion of a disordered feedstock into an ordered product structure reduces entropy, producing waste heat without corresponding energy dissipation.) The conditions for combining reliability and near reversibility are, however, quite stringent: reagent moieties must on encounter have structures favouring the initial structure, then be transformed smoothly into structures that, during separation, favour the product state by ~145 maJ.

The availability of conditional-repetition processes substantially loosens these constraints. A reliable, low-dissipation process can be implemented if the conditions of the preceding paragraph are met, but with a modulation in relative energies not of ~145 maJ, but of ~ (5 + 145/*N*) maJ, where *N* is the maximum number of available repetitions, and 5 maJ is the energy bias favouring the initial structure. Under these conditions, the energy dissipated can be ~1 maJ per operation. For *N* = 10, the required energy difference (~20 maJ) can be induced by relatively modest manipulations of bond strain, overlap energy, electrostatics, or electronic structure. Further, a simple conditional-repetition process with *N* = 10 can achieve high reliability with an exoergicity (and energy dissipation) of ~15 maJ per operation, without need for well-depth manipulations.

In reagent preparation, the feasibility of tailoring the entire reaction environment should commonly permit the use of low-dissipation operations; but in reagent application, the prospects are less clear. Group-transfer reactions often proceed through a single transition state, without the formation and subsequent cleavage of stable bonds linking the reagent device to the workpiece. In reactions of this sort, it should commonly be feasible to choose a group-transfer reagent having bond strengths that make the reaction only moderately exoergic; in addition, it should commonly be feasible to modulate bond strengths so as to permit a close approach to thermodynamic reversibility. In these instances, reagent application causes an energy dissipation of ~15 and ~1 maJ, respectively.

Some reagents, however, form strong bonds to a workpiece without simultaneously sacrificing bonds to their carrier. These must pass though a second, bond-cleaving transition. Whether the energy liberated in the initial bond formation appears as heat or work depends on details of the reaction PES and the reagent device design. In particular, the net exoergicity of bond formation can often be reduced to low values by imposing mechanical constraints that force a temporarily unfavourable geometry on the product (e.g., large angle-bending strains in new single bonds, or large torsional strains in new double bonds). This tends to increase transition state energies, but large piezochemical forces can be used to reduce energy barriers and drive the system along the reaction coordinate. In a suitably designed reagent device, the resulting strain energy can be relieved smoothly as cam-controlled motions modify the geometry of the surrounding structure, thus avoiding the production of thermal vibrational energy. The example of radical coupling (aside from spin-pairing considerations, the time reversal of homolytic bond cleavage). shows that a stiff mechanism can form highly exoergic bonds (~ 500 maJ) without substantial energy dissipation, even without exploiting the strained-intermediate strategy just described.

In light of these diverse strategies for reducing the required reaction exoergicity far below 145 maJ, and for reducing the energy dissipation far below the exoergicity, it seems that mechanosynthetic processes can be quite efficient. In estimating the energy dissipation of systems engineered for good efficiency, it seems reasonable to assume that nearly reversible pathways can be followed in reagent preparation, and that most reagent application steps are either nearly reversible or dissipate an amount of energy characteristic of a moderately exoergic reaction. In this picture, many reactions dissipate ~1 maJ, some dissipate ~15 maJ, and a few (owing to synthetic constraints requiring energetic reagents and precluding energy-recovery strategies) dissipate -100 maJ or more. If the mean energy dissipated per operation is ~ 30 maJ, and the mean number of atoms transferred is ~.1, then the energy dissipated per kilogram of product is ~1.5 × 10^6 J. If this estimate should prove to be wrong, the only conclusions affected are those regarding energy consumption and waste heat generation in manufacturing processes.

Mechanochemical Power Generation

If a mechanochemical process is nearly thermodynamically reversible and is exoergic, then it converts chemical potential energy into mechanical work. Accordingly, mechanochemical systems can be designed to serve as sources of power; via electrostatic generators, they can provide electrical power. (Mechanochemical processes involving charge transfer can provide electrical power more directly, but are not considered here.)

The flexibility of transition-metal chemistry strongly suggests that a wide variety of reactions among small molecules can be carried out in a nearly reversible fashion via intermediate transition-metal complexes, including the reaction

$$2H_2 + O_2 \rightarrow 2H_2O \qquad ...(8)$$

which yields ~475 maJ per molecule of product (delivered to the liquid phase). If conducted at a rate of 10^6 product molecules per second with a mechanochemical system of the size and mass, the resulting power density relative to the mechanochemical apparatus itself is ~8×10^6 W/kg, or ~2×10^9 W/m^3. An overall efficiency >.99 is suggested by the energy dissipation. Similar results can presumably be obtained for the oxidation of small organic molecules.

In molecular manufacturing, the conversion of typical hydrogen-rich, small-molecule feedstocks into diamondoid structures liberates excess hydrogen. If this is combined with atmospheric oxygen, the yield of H_2O typically is on the order of one molecule per carbon atom in the product structure, and the associated net exoergicity is ~400 maJ per atom. This surplus energy can be made to appear as mechanical or electrical power.

Molecular Manipulators

It is sometimes necessary to use a system to build other systems of equal or greater complexity (if this were not done, an infinite regress would make molecular manufacturing impossible). This can be achieved by applying reagent moieties using a multiple-degree-of-freedom positioning mechanism, controlled either by a local nanocomputer or by a more remote macroscale computer: a single mechanism of this kind can perform an arbitrarily complex sequence of operations, with no simple bound on the complexity of the resulting products. Whether directly or indirectly, mill-style systems are likely to be implemented using manipulator-style systems.

Industrial robots are examples of multiple-degree-of-freedom positioning mechanisms. With the availability of nanometer-scale digital logic systems, motors, gears, bearings, and so forth, analogous designs are feasible on a smaller scale (e.g., 100 nm). Aside from differences of scale and component properties, molecular manipulators differ from macroscale devices in that they must maintain positional accuracy despite thermal excitation. This problem can be minimized either (1) by operation at reduced temperatures, which receives no further attention here; or (2) by the use of a stiff mechanism by use of local nonbonded contacts to align reagent devices to workpieces immediately before reaction.

Molecular manipulators typically execute many internal motions (e.g., rotations of drive shafts and gears, displacements of logic rods) per reagent moiety applied. Accordingly, their peak operating frequencies are reduced relative to mill-style systems, and their energy dissipation per operation (at a given frequency) is greater. Peak operating frequencies can exceed 10^6 s^{-1}, although at a substantial (many kT_{300} per operation) penalty in energy consumption. Productivity on a per-unit-mass basis likewise is less than for a mill-style system, but still many orders of magnitude greater than that of conventional macroscale manufacturing systems.

Stiff Manipulator

A sufficiently stiff manipulator can position a reagent moiety relative to a workpiece without needing to exploit interactions with the workpiece to ensure accurate alignment; this simplifies analysis. Completely general positioning requires control of six degrees of freedom (termed six axes): *x*, *y*, and *z*; and roll, pitch, and yaw. These functions could be divided among multiple loosely coupled mechanisms. For example, control of *x* and *y* could be provided by a mechanism moving the workpiece, with control of *z* provided by a mechanism moving the reagent device, and control of the roll, pitch, and yaw orientation of the reagent moiety could be provided by a mechanism that sets and locks the geometry of an adjustable reagent device before loading it into the *z*-axis mechanism. The present section, however, instead describes an integrated, armlike mechanism providing six-axis control. The range of choices in manipulator design is enormous; the specific set of choices explored here is intended to demonstrate lower bounds on capabilities without attempting to present an optimal design. In particular, little effort has been made to minimize energy dissipation, to maximize speed, or to simplify kinematics.

The positioning stiffnesses (for a reagent moiety with respect to a reaction site) of 5 to 10 N/m should be adequate to permit a wide range of reliable synthetic operations. The stiffness of an arm is best analyzed in terms of a series of (additive) compliances. A reasonable design goal is to keep the arm compliance small compared to 0.1 m/

N (e.g., 0.04 m/N), thereby allowing most of the permissible compliance to be allocated to other parts of the system, such as the workpiece and the reagent moiety itself. This goal is, however, not trivial to meet: the scaling law suggests that a 100 nm long arm with a compliance of 0.04 m/N corresponds to a meter long arm with a compliance of 4 × 10^{-9} m/N. This is quite stiff compared to present practice in industrial robotics, motivating a clean-slate design rather than an attempt to scale and adapt existing designs.

General structure and kinematics

A hollow tubular shape can combine low bending compliance with internal space for drive mechanisms. Motion requires joints, which add compliance. A telescoping joint with a threaded interface can combine good structural continuity with adjustable length, adding little compliance. Canted rotary joints (as in some experimental robot arms and hard-shell space suits) can permit angular motion of the arm with respect to the base and of the working end with respect to the arm, while again maintaining good structural continuity. The helical threads of the telescoping joint and the circular threads in the bearings of the rotary joints can form high-shear-stiffness interfaces. The external form

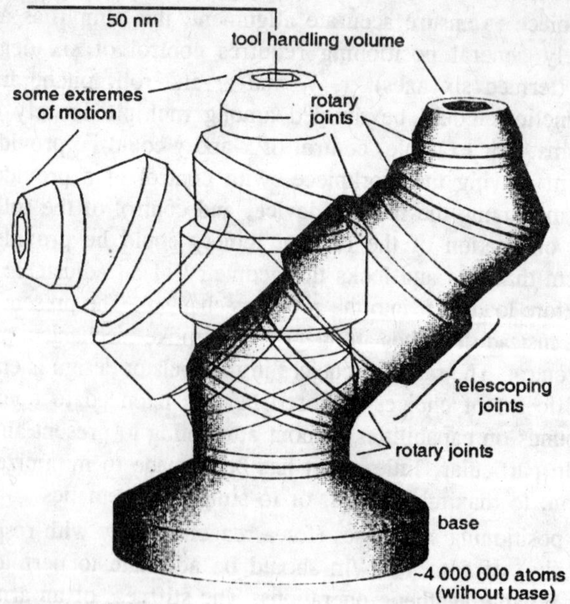

Fig. 5.9. External shape and range of motion of a stiff arm design compatible with implementation on a 100 nm scale.

and range of motion of a device with these characteristics; it has a maximum extended length (with tool holder) of 100 nm, a diameter of 30 nm, and a typical wall thickness of 3.5 nm.

A drive system forces the tubular structural segments to rotate with respect to one another, and thus to rotate with respect to the local axis of the arm. In some instances these rotational motions must be accomplished while maintaining high torsional stiffness. The approach chosen here exploits toroidal worm drives driven by mechanisms in a flexible core structure.

A stiff arm of this design has substantial dynamic friction, yielding relatively poor trade-offs between speed and energy dissipation. Speed and efficiency can be substantially improved if the arm need not move through a large distance between operations. Accordingly, the present design reserves a substantial volume of space for a transportation pathway from the base to the tip, permitting reagent moieties to be interchanged without arm motion by means of an internal conveyor mechanism.

Power supply and control

The power supply and control system is outside the arm itself, and hence is not subject to tight geometric constraints. It need only be assumed that shaft power is available, and that a control mechanism can engage and disengage smaller drive shafts (using clutches), causing them to turn by a set number of rotations between locked states. Reasonable implementations can be devised that use gear trains to produce an odometer-like encoding of the number of shaft rotations and use a corresponding device to encode a representation of the desired number of shaft rotations (this device can be written to like a register); a match between the two can be made to trigger a motion causing the disengagement of a clutch and the locking of the corresponding drive shaft. To initiate a motion, a digital logic device stores a suitable number in the register, then engages a clutch to cause shaft rotation in the appropriate direction.

Drive shafts undergo longitudinal displacements as the arm moves. A coupling that permits these motions while applying torque (e.g., a drive nut engaging longitudinal grooves on a shaft segment) can interface these shafts to drive and control mechanisms with fixed positions.

Drive shafts, gears, and the core structure

Fig. 5.10 shows a cross section of an arm, including all the moving parts required for arm motion save for three drive shafts (and gears)

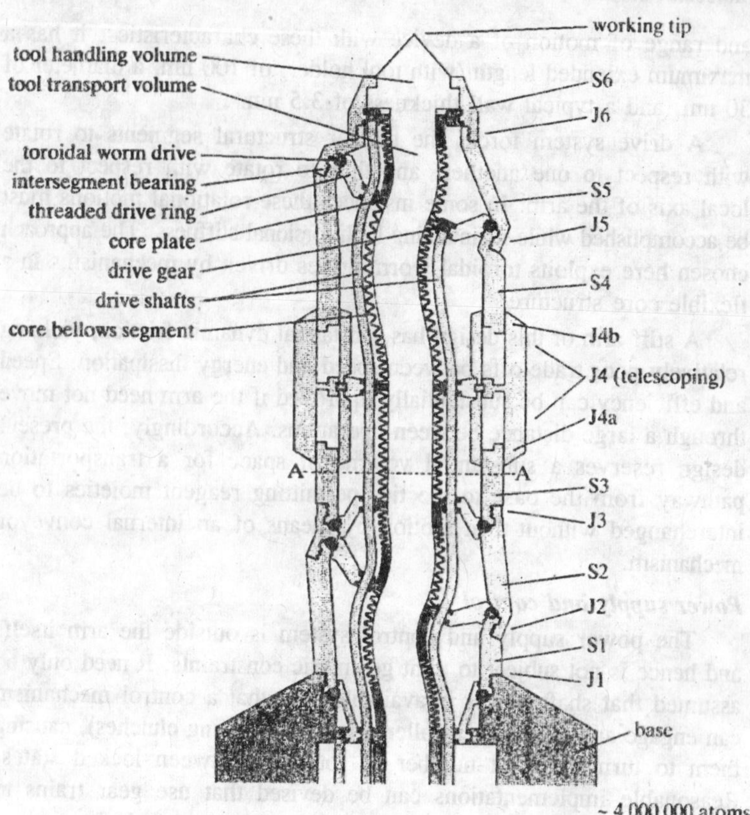

Fig. 5.10. Cross section of a stiff manipulator arm and identification of parts.

not in the plane of the section. As shown, the shafts have diameters of 1.5 nm; the compliance analysis assumes the use of a material with a modulus ~0.1 that of diamond, lowering the stresses caused by shaft bending. Surface strains in the shafts are well within the acceptable range even of high-modulus diamondoid structures. One shaft, ending at joint J1, need not be flexible and is located outside the core structure.

The remaining shafts are threaded through bearing apertures in *core plates*, and are free to rotate and to slide longitudinally with respect to these plates. The rims of the core plates interlock with drive rings that in turn interlock with joints in the arm, forcing coplanarity between each core plate and its corresponding joint. Both segments linked by the joint can rotate with respect to the core plate. Core plates are linked by core bellows segments. These flex to

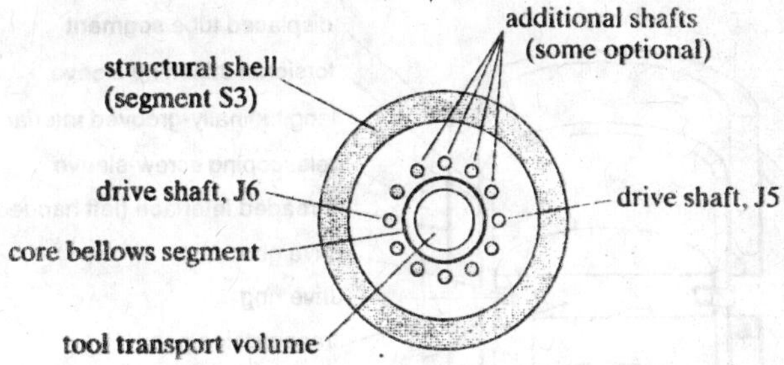

Fig. 5.11. Cross section A—A of arm in figure 5.10, perpendicular to the axis.

accommodate the angle and offset between the planes of the joints; the angle and offset have fixed magnitudes and relative orientation, but they can rotate with respect to the core structure. The core bellows segments link the core plates in a manner that gives the core structure substantial torsional stiffness. (Metal bellows often are used to transmit torque between misaligned shafts in macroscale engineering practice, despite the lower elastic limit and susceptibility to fatigue damage characteristic of macroscale metal structures.) Additional structures resembling the core plates could be introduced to further constrain and shape the paths of the bellows and drive shafts.

Within the envelope of the core bellows segments and core plates is a region with a diameter of 7 nm that plays no role in the arm structure or in driving its motions. This region is reserved for a mechanism to transport reagent moieties (or reactive clusters) to the tip of the arm, where a larger region is reserved for a mechanism to position and lock these moieties or clusters into position for use. An articulated conveyor system based on belts, rollers, and sliding-interface guideways can serve the transport function. Paths like those allocated for drive shafts can carry several shafts for actuating the reagent positioning and locking mechanism, or (in related applications) for transmitting signals from a sensor at the tip of the arm to a computational system outside the arm.

Drive shafts end in drive gears that engage the inner surface of a drive ring. As a drive shaft rotates, the drive gear forces rotation of the drive ring with respect to the core plate, which is itself restrained from substantial rotation about its axis by the torsional stiffness of the core bellows segments.

Joints and joint drive mechanisms

The internal structure of the arm design including the components of the six joints. The joint nearest the tip of the arm, J6, plays no

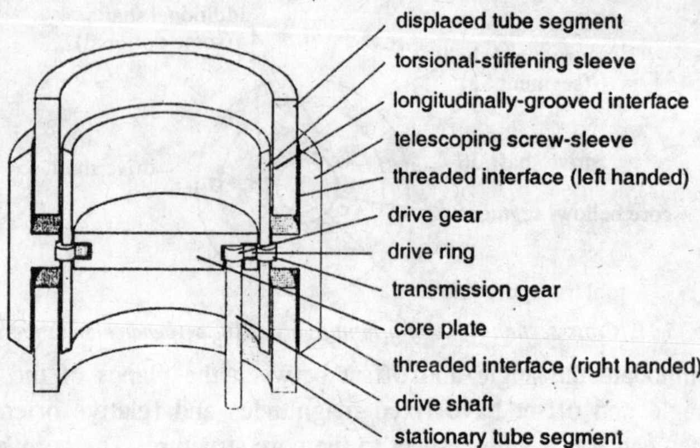

Fig. 5.12. Structure and kinematics of a stiff telescoping joint.

role in positioning and transverse compliance, affecting only the rotational angle of the tool. Its stiffness requirements are slight, hence it can be directly geared to a drive shaft.

The other rotary joints, J1, J2, 13, and J5 are implemented as toroidal worm drives. Harmonic drives offer an alternative to the toroidal worm drive, but if repulsive interactions are used to implement their teeth, the large and time-varying radial forces applied by the wave generator are likely to result in larger energy dissipation at the interface between the wave generator and the flexspline than at the corresponding interface between the threaded drive tube and the triple-threaded torus. Further, the large asymmetrical loads would require additional structural analysis.

The telescoping joint assembly, J4 (including threaded interfaces J4a and J4b*)*, can be directly geared to a drive shaft, since the thread angles (with a thread pitch of 0.5 nm) provide a speed reduction that provides ample stiffness.

Estimated arm compliance

The compliance of the arm (as in most beamlike structures) is greatest in bending modes. This section will estimate compliance in arms with the tip pointing upward or inward in the worst-case geometry of full J4 extension and maximum bending. The compliance is then greatest for displacements perpendicular to the direction of bending. Table 5.1 summarizes the main contributions to total arm compliance in this geometry, and the physical assumptions behind the calculated contributions from structures outside the core and drive system. The compliance contributed by the core and drive system can be reduced

Table 5.1. Major contributions to transverse compliance of a manipulator.

Source	*Coordinate*	*Magnitude (m/mN)*	*Multiplier*	*Contribution (m/mN)*
Tube bending	End displacement	13.60	1	13.6
J1 rockint	Max. local stretch	0.05	59	3.0
J2 rocking	Max. local stretch	0.05	45	2.3
J3 rocking	Max. local stretch	0.05	32	1.6
J4a rocking	Max. local stretch	0.05	19	1.0
J4b rocking	Max. local stretch	0.05	9	0.5
J5 rocking	Max. local stretch	0.05	4	0.2
J6 rocking	Max. local stretch	0.07	3	0.2
J1 torsion	Max. local shear	0.31	25	7.8
J2 torsion	Max. local shear	0.31	7	2.2
J3 torsion	Max. local shear	0.31	1	0.3
J5 torsion	Max. local shear	0.31	0.5	0.2
SI torsion	Max. local shear	0.11	25	2.8
Drive system				4.0
Total				<40.0

by using toroidal worm drives having large speed-reduction ratios. The compliance of the arm can be made almost independent of the core and drive compliances by exploiting the relationship

$$C_{s,2} = (v_2 / v_1)^2 C_{s,1} \qquad ...(9)$$

where $C_{s,1}$ and $C_{s,2}$ are the compliance of the system measured with respect to displacements at points 1 and 2, and v_1 and v_2 are the speeds of those points as the system executes some motion. (This expression assumes no compliance between points 1 and 2, i.e., that their velocities and displacements remain in a fixed ratio.) A mechanism providing a large speed-reduction ratio between the motion of the surface of the drive gear and the motion of the driven segment can tolerate large drive-shaft and power-supply compliances.

Drive shafts with a diameter of 1.5 nm and a shear modulus of 5 $\times$ 10^{10} N/m^2 will have a torsional compliance (for surface displacements) of ~6 m/N at a distance of 100 nm from a torsionally stiff locking mechanism. Torsional compliance in the core bellows structure adds directly to the drive rod compliance, but this increment can be made comparatively negligible. For J6, direct gearing of the joint to the drive gear gives a standard deviation in joint angle of

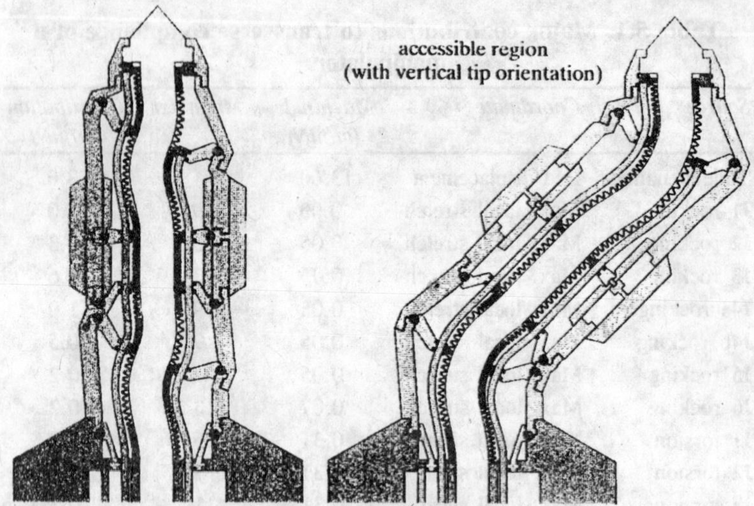

Fig. 5.13. Cross section of a stiff manipulator arm.

~0.02 radians. This is orthogonal to the other compliances discussed here, and is negligibly small for most mechanochemical operations involving small reagent moieties.

For J4, if a drive gear is coupled to rotation of the threaded outer sleeve with a gear ratio of unity, it contributes a compliance in manipulator extension of ~0.018 m/N, assuming a threaded interface with a radius of 15 nm and threads of 0.5 nm pitch. This is orthogonal to the bending compliances considered elsewhere in this section.

In J2, J3, and J5, drive-rod compliance contributes to transverse-bending compliance. In each, the difference in radius between the geared inner surface and threaded outer surface of the drive ring contributes a speed increase of ~1.8, ~1.5, and ~1.8 respectively. The respective drive-rod lengths are ~25, ~35, and ~75 nm. The corresponding contribution of each of these joints to the transverse compliance at the tip can be limited to ≤0.001 m/N apiece if the speed-reduction ratios of the toroidal drive assemblies are 190, 70, and 65, respectively, measuring ratios from the speed across the threaded interface of the drive ring to the speed of the driven segment. (Toroidal worm drives of this size can provide ratios >3000.) In turning a tube segment through one full rotation, even the J3 and J5 drive shafts undergo > 700 rotations.

For J1, no flexible drive rods are needed, hence the drive system compliance can be limited to a small value. Taking 0.1 m/N as a readily achievable target, the contribution to the tip compliance can be limited to ≤ 0.001 m/N with a speed-reduction ratio of 50. Drive system contributions to transverse compliance then total ≤ 0.004 m/N.

The total arm compliance estimated is 0.04 m/N, corresponding to a stiffness of 25 N/m. This permits the use of moderately compliant workpieces and reagent moieties (~0.1 m/N) while keeping error rates $\leq 10^{-15}$ in directing reactions to chemically equivalent sites separated by a single bond length. This stiffness also permits the application of 1 nN forces with an elastic deflection in the arm of only 0.04 nm. (Elastic deflection under a planned load need not contribute to positional error.)

Speed, productivity, and magnitude of power dissipation

To estimate the power dissipation of the manipulator system on a per-operation basis requires a model of a typical motion, which in turn requires consideration of a typical sequence of operations in the synthesis of diamondoid structures. To build such a model, let us assume a synthetic strategy based on the transfer of small moieties to a surface. The characteristic scale length of required motions can be taken as the 0.25 nm spacing of the bonds perpendicular to a diamond (111) surface. Efficiency favours small motions between operations. Consider an operation at a typical point (a), with subsequent operations required both at a set of ~6 neighbouring points (b) and at a set of more distant points (c). After completing an operation at point (a), a large motion would be required if and only if synthetic constraints somehow demanded that each of the operations in set (b) be delayed until after executing an operation at a point (d) in set (c). Save in rare or contrived circumstances, this somewhat unusual constraint could arise only if the point (d) were adjacent to a point in set (b), thereby permitting a substantial electronic or steric interaction between them. Accordingly, the required motion distance should often be one scale-length (0.25 nm), occasionally be two (0.5 nm), and only rarely be greater. Treating 0.5 nm as a typical displacement distance between operations should yield conservative estimates of energy dissipation.

The present arm design, however, causes substantial tip displacement in the course of executing a rotation in joints J3 and J5, which are provided chiefly to control tip orientation. Assuming that such motions are commonplace and result in tip displacements of ~10 nm, it is conservative to assume that joints J1 and J2 will commonly have to execute motions able to cause comparable but compensating tip displacements. If such motions are performed in ~10^{-6} s, the speed of the fastest sliding motion (that of the J2 drive ring) is ~1 m/s. Assuming that phonon drag at sliding interfaces (which scales with speed squared) is the dominant power-dissipation mechanism, motion in this joint can be expected to account for a substantial fraction of the total power dissipated.

To limit shear displacements of a threaded drive ring with respect to the triple-threaded torus of the worm drive, an interfacial shear stiffness of ~100 N/m is ample; the interfacial compressive stiffness can be several times this value. The corresponding values of the phonon transmission coefficient, T_{trans} are ~10^{-4}, (assuming diamondlike materials). Applying assumptions like those used in the bearing analyses to a sliding-interface area of ~200 nm^2 yields an estimated energy dissipation per motion of ~10 maJ. An estimate of ~100 maJ per motion for the arm as a whole therefore appears generous. Note, however, that this estimate does not include energy dissipated in the power and control system, and is rather crude.

If an arm emplaces ~1 atom per motion, a dissipation of ~100 maJ is significant on the chemical energy scale. If an arm emplaces a cluster of ~2 nm size, containing ~1000 atoms, the energy dissipated per atom is a trivial ~0.1 maJ.

These calculations indicate that arms performing 10^6 operations per second need not incur excessive penalties from friction (although the energy cost is large compared to that of a mill-style mechanism). The total number of atoms in a mechanism of this size is ~5×10^6 (neglecting the base, power and control system, etc.); accordingly, the time required for a mechanism of this type to perform the motions necessary to build a product structure of similar mass and complexity is ~ 5 s. This suggests that the use of manipulator mechanisms need not preclude high overall system productivity.

Toward better designs

The design is far from optimal. Although maximizing stiffness is a major design objective, and the major sources of compliance are in the region from joints J1 to J3, these joints and tube segments are no stiffer than the less critical structures further along the arm. A design based on a Stewart platform might provide superior stiffness.

Minimizing energy dissipation is another major design objective. Increasing joint stiffness and using several drive shafts for critical joints can lower the magnitude of required speed-reduction ratios, decreasing the sliding speeds in interfaces and lowering dissipation. Other major improvements can doubtless be made. The present design exercise merely shows that suboptimal designs are adequate to enable reliable, reasonably efficient molecular manipulation.

Compliant Manipulators

A stiff manipulator can be used to apply small, simple reagent devices to workpieces without any special adaptation to the workpiece structure (aside from choice of moiety, position, orientation, force, etc.). At the cost of greater tip complexity, reactive encounters can exploit nonbonded interactions to achieve reliable alignment, permitting substantially greater arm compliance while maintaining high reaction reliability.

The intermediate states in the construction of a diamondoid structure can be limited to surface irregularities of atomic height on a nanometer length scale in the vicinity of each reaction site. To perform a reaction at such a site, a tip can be configured to have regions of specific steric complementarity with the workpiece surface, bringing the tip into stiff alignment (compliance ≤0.01 m/N) with the surface before the final reactive encounter. The required control of tip geometry can be achieved either by adding adjustable probes to the end of the arm (e.g., configured by geared mechanisms driven by rotating shafts), or

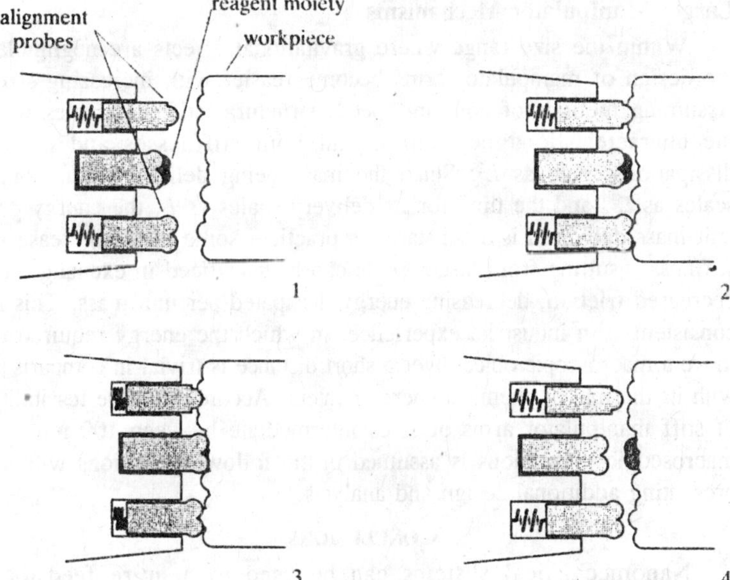

Fig. 5.14. Schematic diagram showing alignment of a reagent moiety with respect to a workpiece aided by sterically complementary probes.

by attaching probe structures of suitable geometry (selected from a diverse set) to the reagent device beforehand. Multiple-position detent mechanisms can be exploited to provide the necessary steric diversity.

Error Rates and Sensitivities

Manipulator-style mechanisms will tend to have lower stiffness and hence higher error rates than those of mill-style mechanisms. Nonetheless, the feasible stiffness permits errors stemming from thermal excitation to be kept to negligible levels ($\ll 10^{-15}$), given suitable choices of reagent, reaction site, and encounter geometry. Self-aligning tips can increase stiffness, broadening the range of arm structures and reaction conditions that are compatible with low error rates.

Manipulators under programmable control can more easily be designed for fault tolerance than can mill systems. With the ability to vary the sequence of reagent devices and motions, and with the ability to make these sequences contingent on the results of measurements of the workpiece structure, it is feasible to continue operations after discarding damaged workpieces and tools. In some instances, flaws could be corrected, but this capability is not assumed in the following analyses.

Larger Manipulator Mechanisms

Within the size range where gravitational effects are negligible, the design of manipulator arms becomes easier with increasing size. Assuming motions of constant speed, structural stiffness scales with the characteristic dimension L, and joint stiffnesses and power dissipations scale as L^2. Since the mass being delivered commonly scales as L^3, and the time for its delivery scales as L, the energy per unit mass delivered is a constant. In practice, some of the increase in stiffness resulting from larger scale can be sacrificed in exchange for decreased friction, decreasing energy dissipated per unit mass. This is consistent with industrial experience, in which the energy required to move a macroscopic object over a short distance is trivial in comparison with its thermal or chemical energy content. Accordingly, the feasibility of stiff manipulator arms of sizes intermediate between 100 nm and macroscopic dimensions is assumed in the following sections without presenting additional design and analysis.

CONCLUSIONS

Nanomechanical systems can be used to acquire feedstock molecules from solution, sorting and ordering them to provide a flow of input materials to essentially deterministic molecule-processing systems. Exploiting mechanochemical processes, these systems can convert feedstock molecules into reactive moieties of the sorts. These moieties, in turn, can be applied to workpieces in complex patterns to build up complex structures.

Each of these operations can be made sufficiently tolerant of thermal excitation that damage to manufacturing mechanisms and product structures occurs chiefly as a result of ambient ionizing radiation. Molecular sorting operations can be quite efficient, with energy dissipation <1 maJ per molecule. Assembly operations using mills can be energetically efficient, with the energy dissipated in many operations being <1 maJ, and an estimated mean dissipation for a mix of efficient and inefficient operations being ~ 30 maJ per operation. The efficiency of manipulator motions is expected to depend heavily on the energy dissipated by their control mechanisms.

Each system has been characterized at a frequency of 10^6 operations per second. At this frequency, the sorting, processing, and manipulation subsystems each process a quantity of reagent moieties equaling its own mass within a few seconds. The next chapter describes how fast, reliable processing units like these can be combined with larger-scale assembly systems to build molecular manufacturing systems capable of delivering macroscopic products.

Some open problems

As in the previous three chapters, the task of expanding the range of well-characterized, atomically detailed models for devices defines a large set of open problems. Modulated receptors can readily be designed within the constraints of existing software and computers. A more detailed understanding of purification processes could be developed by choosing a small feedstock molecule, compiling an exhaustive list of similar molecules of smaller size, and characterizing the interactions of each with feedstock receptor structures of several designs.

Many of the problems of greatest interest in mill and manipulator design depend on the prior characterization of a compatible set of mechanochemical processes; the discussion in this chapter highlights the interest in identifying sets of nearly thermodynamically reversible processes that can combine fuel and oxygen molecules. Various mechanical components of mill and manipulator systems (providing transport, power supply, alignment, error detection) are independent of the specifics of the mechanochemical process being performed, hence their detailed design can proceed with existing knowledge (subject, as always, to limitations in the accuracy of molecular mechanics methods).

6

Molecular Manufacturing

Several considerations (assembly speed, the mass of products in transit, the size of assembly units in fault-tolerant architectures) favour the use of convergent assembly over moiety-by-moiety assembly in making product structures of substantial size and complexity. In convergent assembly, small parts (initially individual transferable moieties) are combined to make larger parts, and these to make still larger parts, in a hierarchically organized process. Convergent assembly is familiar in macro scale manufacturing processes as well: in making an automobile, parts are combined to make engines before engines are joined to the chassis; electronic components are combined to make circuit boards before boards are installed into radios, and radios are assembled before being joined to the dashboard.

Assembly Operations at Intermediate Scales

Joining Building Blocks

Positioning

Both mill-style and manipulator-style mechanisms can be used to transport and place building blocks of intermediate size. Manipulators of sizes intermediate between the 100 nm and macroscopic scales; systems of belts, rollers, and cams can also be built at intermediate scales, again providing increases in mechanical stiffness.

The scale of these mechanisms depends on the scale of the parts being handled and of the product being assembled. In mill-style mechanisms, the transport system components must be scaled to the input parts. In manipulator-style mechanisms, the range of motion and hence the size of the device must be scaled to the output product.

Some useful assembly operations at intermediate scales may be quite specialized, such as stretching the sleeve of a grooved sleeve bearing in order to fit it over a shaft. Although specialized mechanisms are not explicitly considered here, they are likely to be commonplace in practice.

Interfaces between blocks

The assembly of blocks to form larger structures spans a continuum of sizes between cluster-based synthesis strategies and joining of macroscopic blocks. At sufficiently large scales (e.g., ~100 nm) the use of snaps, screws, and the like is feasible. On all scales above the smallest, adhesive interfaces of the sort are applicable. As noted, some adhesive interfaces can join to form a dense, continuous diamondoid structure (or diamond itself). As a consequence, designing systems to be built as compositions of many small parts need not entail any great sacrifice in structural strength and stiffness.

Systems of molecular machinery will commonly consist of many distinct, free-standing parts, joined by interlocking geometries rather than by covalent bonds. Systems of this sort can be built by methods resembling conventional robotic assembly, using grippers that do not bond to the part being moved. Surface forces (e.g., van der Waals attraction) must be considered in designing grippers and assembly sequences, but these present no great difficulty where diamondoid structures with nonreactive surfaces are concerned, given assembly adhesion. (The construction of free-standing parts can be accomplished by building up a workpiece on a substrate in a manner that allows clean cleavage at the workpiece-substrate interface; the cleaved surface can then be finished or extended by further mechanosynthetic operations. The forces required for cleavage and subsequent manipulation can be provided by nonbonding grippers.)

Energy efficiency

The energy efficiency of intermediate-scale assembly can be high, on a per-atom basis. In moving large blocks, internal friction in positioning mechanisms is amortized over many atoms. A typical upper bound on the energy dissipated in joining surfaces is comparable to the surface energy, which for a large, thick block is small on a per-atom basis.

Energy dissipation is accordingly dominated by assembly of the smallest blocks. These can be approximated as one-nanometer cubes. If the energy dissipated in joining surfaces were equal to the theoretical surface energy of diamond (~5 J/m^2), this would amount to ~9 ×

10^6 J/kg. It is reasonable to expect, however, that much or almost all of the potential energy of bonding can be recovered as mechanical work, reducing dissipation to a small fraction of this value. If adhesive interfaces were designed such that the exoergicity of C—C bond formation matches that in the polymerization of ethene, then the exoergicity of assembling a diamondoid solid from cubic nanometer blocks would be $\sim 1 \times 10^6$ J/kg. This value probably overestimates the total energy dissipated in block assembly in a well-designed system producing a mix of diamondoid products; the estimate used here is 5×10^5 J/kg.

Reliability Issues

In the assembly of intermediate-scale building blocks, ample exoergicity for stable, reliable assembly is no problem; even van der Waals attractions yield > 150 maJ for typical interfaces of ≥ 1 nm^2. Further, the greater stiffness of larger structures makes it increasingly easy to limit positional uncertainty resulting from thermal excitation to low, acceptable values. The special problems arising on a larger scale are greater sensitivity to nonthermal vibration resulting from an increased ratio of mass to stiffness, and greater radiation damage rates resulting from increased mass directly.

Vibration and alignment

Under a fixed acceleration, inertial loads cause elastic deformations that scale in proportion to the square of the characteristic length of a system. Oscillating accelerations (vibrations) produce dynamic deformations that depend on resonant frequencies and damping. For perspective, a manipulator 10 cm long and scaled from the device has a length and stiffness greater by a factor of 10^6 and a mass greater by a factor of 10^{18}. It deforms by ~ 10 nm under terrestrial gravity, and its lowest resonant frequency is $> 10^4$ Hz.

Various combinations of (1) assembly tolerances, (2) measurement and control, and (3) precomputed corrections can compensate for static deformation. Various combinations of (1), (2), and mechanical isolation can correct for vibration, under ordinary circumstances. A surprise in the development of scanning tunneling microscope technology was the feasibility of achieving atomic-scale positional stability in centimeter-scale mechanisms mounted on ordinary laboratory tables, using 10-cm-scale vibration isolation systems consisting of ordinary elastomers and metal plates. Engineering molecular manufacturing systems for isolation from external vibrations and for control of internal vibrations should be feasible.

At larger scales, assembly tolerances can be increased without sacrificing precise control of the structure and alignment of the resulting interfaces. An adhesive interface structure in which nonbonded contacts occur as tapered pegs are inserted into matching holes. Tapering can provide substantial tolerance for initial misalignment, yet as the adhesive surfaces themselves approach contact, a stiff alignment can be achieved. As shown in the illustration, interfacial bonding can proceed outward from the alignment pegs, and the structure around the edge of the nonbonded contact surface can be placed in compression to inhibit crack initiation.

Radiation damage

Typical nanometer-scale machine components have many locations where a point defect induced by ionizing radiation can cause failure; the default assumption throughout this work has been that a defect at any point will cause failure. Macroscale machines, however, function despite numerous defects. At some intermediate scale, machines can (and must) be designed such that a significant density of point defects is tolerable. This scale can be estimated by considering the effects of radiation damage on two basic characteristics of mechanical parts: their shapes and the properties of their interfaces. (Strength, stiffness, and the like are not at issue in the typical case.)

Applying an elastic continuum model, the effect of a point defect on the shape of a component can be described in terms of the resulting strain field. This is in general complex, but its longest-range component results from the local change in volume. Modeling the medium as homogeneous, isotropic, and elastic, the resulting radial strain Δr falls off as r^{-2}. An atomic-scale defect causes a strain of $\Delta r < 0.2$ nm at $r = 0.2$ nm, and hence a strain $\Delta r < 0.002$ nm at 2 nm, and so forth. Since geometrical errors of ~ 0.01 nm are tolerable in most mechanosynthetic systems, components having features of multinanometer or larger scale should usually prove able to tolerate radiation-induced defects in their interiors without failure. Large structural components with damage-sensitive interfaces can have substantial damage-tolerant interior and surface regions, in effect reducing their target sizes relative to the model.

Similar principles apply to misalignments caused by damage to bearing interfaces, although the nonlinearity of nonbonded interactions tends to increase the characteristic length scales. Moreover, damage occurring at a mobile interface can in some instances produce large static friction or adhesion, causing device failure. Bearings can be made damage tolerant by providing internal redundancy. For example, the surfaces of a sleeve bearing can be separated by multiple independent layers, each subdivided into parallel strips. A structure of this sort can tolerate several adhesion points among adjacent surfaces, strips, and layers without creating a connection between the sliding surfaces. Designs of this sort typically require interfacial thicknesses of several nanometers, which are geometrically compatible only with moving parts on a scale of tens of nanometers or more. Thus, systems large enough to employ internally redundant bearing interface structures commonly have structural components large enough to maintain good geometrical stability in the presence of point defects.

In a mill-style manufacturing process without fault-correction, a failed assembly operation can cause failure of a relatively large subsystem. The estimates made in this section regarding the scale at which machine components can tolerate point defects without failure suggest that assembly processes can tolerate damage in the parts themselves when these are of a similar scale. In the presence of point defects, assembly can proceed without disrupting the process; whether the resulting product is functional is a separate question. Accordingly, it will be assumed that when the scale of all components and building blocks in a manufacturing process is > 10 nm, a moderate density of point defects can be tolerated without causing failure of the process.

Architectural Issues

The previous sections of this chapter have chiefly described devices, subsystems, and operations, covering molecular sorting and orientation, transformation of molecules into reagent moieties, the application of reagent moieties to build small structures, and the hierarchical assembly of smaller structures into progressively larger structures. The present section outlines an architecture for molecular manufacturing systems capable of producing a wide range of macroscopic products and delivering them to an external environment. This requires attention to the spatial organization of material and energy flows, and to such system-level concerns as overall reliability and operational lifetime. These are addressed in substantial detail in the description of an exemplar architecture.

Combining Parts to Make Large Systems

Present manufacturing practice provides no precedent for processes that combine $\sim 10^{25}$ distinct parts to form a single object. A naive approach would use a mechanism to assemble parts one at a time; at 10^6 per second, however, assembling 10^{25} parts would take a time longer than standard estimates of the age of the universe. Viable approaches must coordinate many mechanisms in building a single object.

Construction-style assembly processes

A construction-style process uses small devices to work on or in a large structure. For example, many assembly mechanisms can work in parallel to build up a surface. If a product structure is 0.1 m thick, and each assembly mechanism occupies an area of $(100\ nm)^2$, then an assembly rate of 10^6 small moieties per second implies a $\sim 10^8$ s construction time, more than a year. This strategy might be attractive for designs that lack thick solid regions: various porous, fibrous, foamlike, or honeycombed structures meet this criterion and have attractive mechanical properties.

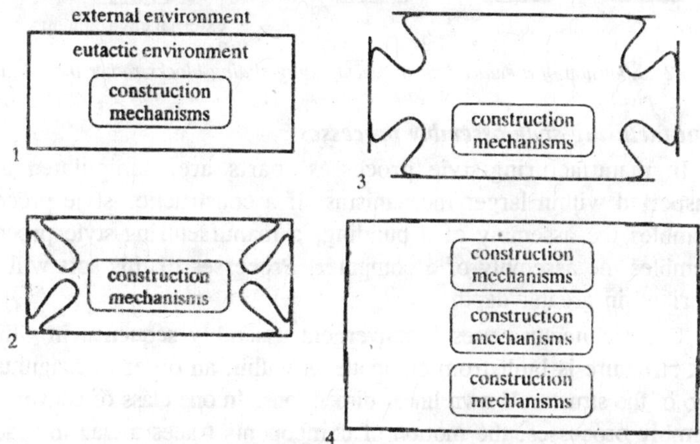

Fig. 6.1. Expansion of a eutactic environment by construction mechanisms operating inside it.

Alternatively, assembly can proceed at higher rates through the use of larger building blocks. Keeping other parameters constant, increasing the block size to ~ 10 nm decreases the assembly time to 1000 s. Blocks of sufficient size and complexity (e.g., containing a computer, motor, and actuators) could be made self-assembling, although at a substantial penalty in properties such as strength-to-density ratio.

Some products can doubtless be assembled in a poorly controlled environment (e.g., in a solvent bath), simplifying problems of heat and mass transport, and of environmental control. Alternatively, eutactic environments of almost any desired size can be constructed by expanding a gas-tight barrier.

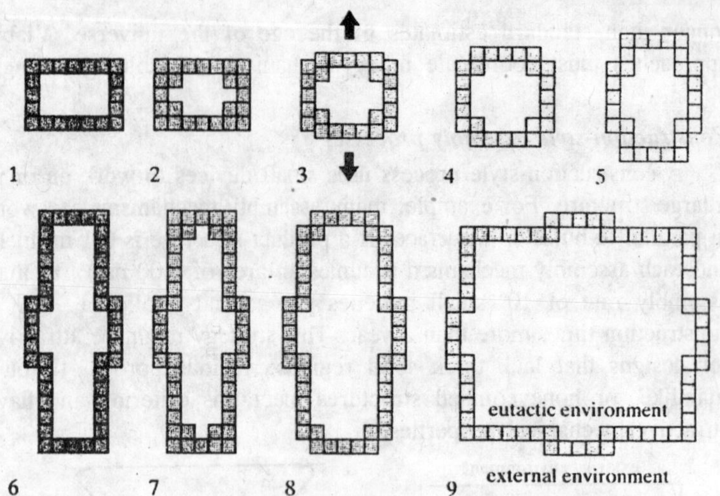

Fig. 6.2. Expansion of a eutactic environment, using sliding blocks rather than flexible pleats.

Manufacturing-style assembly processes

In manufacturing-style processes, parts are manipulated and transported within larger mechanisms. If a construction-style process resembles the assembly of a building, a manufacturing-style process resembles the assembly of a computer. Processes of this sort will be described in greater depth.

The architecture uses a convergent assembly sequence in which each structure is built from components within an order of magnitude or so of the structure's own linear dimensions. In one class of convergent assembly processes, the motion of components traces a tree in space: the trunk corresponds to the path traced by the final workpiece as the final components are assembled, the branches correspond to the paths traced by those components as they are assembled, and so forth.

A simplified model demonstrates that convergent assembly can be distributed in space in a manner that (1) provides an assembly volume proportional in size to the workpiece at each stage, and (2) requires only short-range transportation of parts between stages. A system of the sort must operate all assembly stations at the same frequency, thereby slowing the smallest mechanisms to the frequencies of the largest.

Varying the number of assembly operations per level in the hierarchy can alleviate this problem, as can allowing a small number of highfrequency processes to supply parts to a larger number of low-frequency processes. Both of these strategies are incorporated into the exemplar architecture. The boundary between high-frequency and low-frequency processes in the exemplar architecture corresponds to a boundary between mill-style and manipulator-style assembly mechanisms. This, in turn, marks the boundary between purely repetitive operations producing standard building blocks and programmable operations that can stack these building blocks to make a wide variety of products. Although manipulators can handle blocks of ~ 10 nm scale with good energy efficiency, the exemplar architecture uses mill-style mechanisms up to a 1μ scale. This sacrifices flexibility, but the number of distinct kinds of 1μ blocks can be quite large. A 1 kg structure contains $\sim 10^{15}$ blocks. To manufacture them, the exemplar architecture uses $\sim 10^{6}$ separate systems, each capable of producing a different structure. The demand for large numbers of a few kinds tends to reduce diversity, but the option of running some lines at low capacity (or only for producing certain products) substantially offsets this. The range of systems that can be constructed from a catalogue of $\sim 10^{6}$ kinds of parts is large, particularly when the parts can be joined to make continuous volumes of high strength materials, and are each smaller than the usual tolerances in conventional manufacturing.

Delivering Products to an External Environment

In typical applications, the products of molecular manufacturing must be delivered to an external environment without permitting back contamination of the eutactic internal environment. For example, a product object can be placed in a box bearing multiple external sealing rings and followed by a similarly ringed piston. The rings can provide a tight seal against the inside of an exit tube, and the piston can block

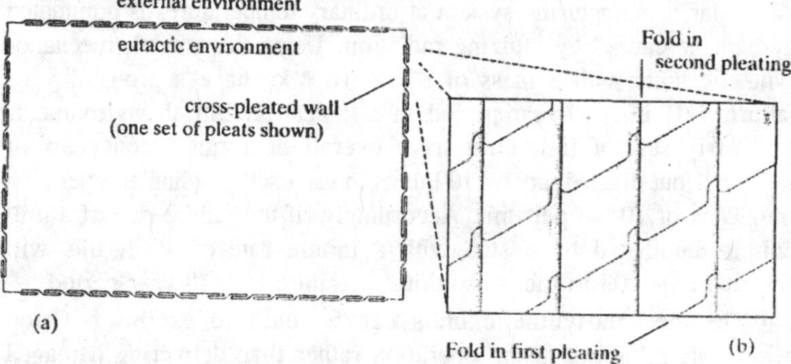

Fig. 6.3. Schematic illustration of a gas-tight, expandable, doubly pleated enclosure for a eutactic environment.

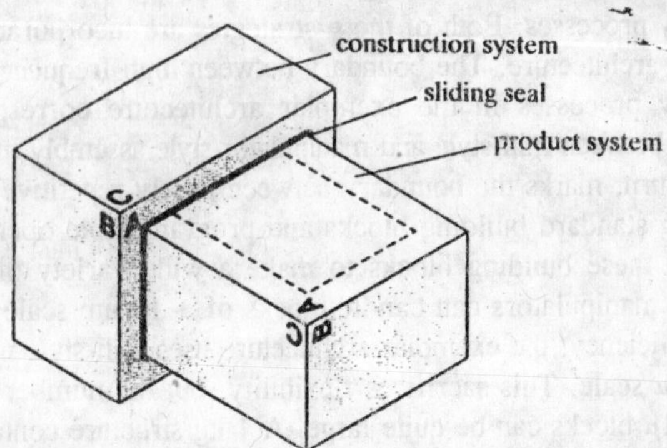

Fig. 6.4. Schematic illustration of a geometry permitting the manufacture and delivery of objects of a size equaling or exceeding that of the manufacturing system.

the exit until another product-bearing box is ready for departure. Moderate failures in the seals could be rendered harmless by scavenging pumps operating on the spaces between them. Schemes that lack discard able boxes, pistons, and the like can also be devised, with varying volume requirements, constraints on product object shape, and so forth. These will not be pursued further at present.

Redundancy, Reliability, and System Lifetimes

A system of macroscopic scale and significant lifetime must tolerate localized, randomly distributed damage, which in nanomechanical systems will commonly cause failure of devices and subsystems. There is a large and diverse literature on fault-tolerant design for computers and aerospace systems, and similar approaches are applicable in many nanomechanical systems.

With a judicious choice of structures and reactions, damage to a molecular manufacturing system at ordinary temperatures is dominated by damage caused by ionizing radiation. Using this model, mechanosynthetic units with a mass of $\leq 2 \times 10^{-18}$ kg have a probability of failure $\leq .01$ in a 10-year period in a typical terrestrial environment. The early sets of mill units have overall operating frequencies of $\sim 10^{6}$ s^{-1}, but divided among 100 units in each set, yielding an operating frequency of 10^{4} s^{-1} per unit. Accordingly if the failure rate of a mill unit is dominated by a step with a failure rate of 10^{-15}, this will contribute $\sim .003$ to the probability of failure in a 10-year period.

The mechanosynthetic units can be built to exhibit fail-stop behaviour, reliably ceasing operation rather than delivering damaged products. The characteristics of transportation components capable of transferring structures from belts to pallets, moving pallets along tracks

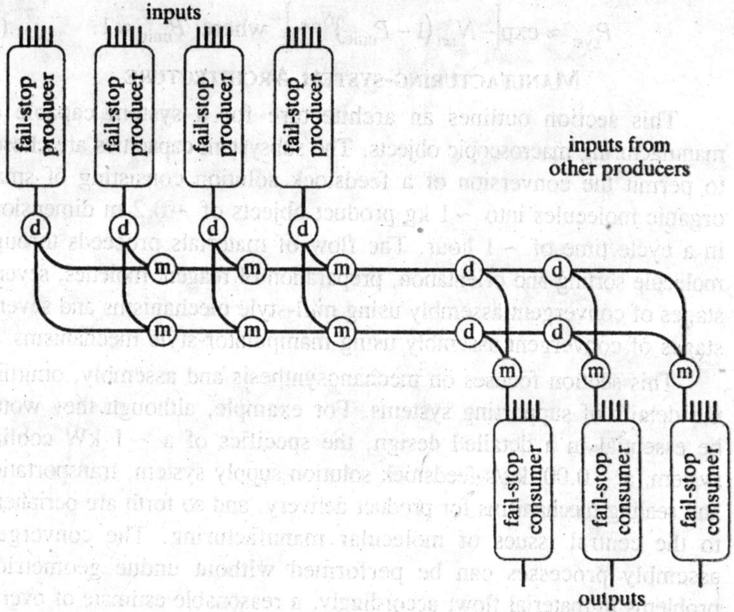

Fig. 6.5. Schematic diagram of a portion of a redundant, fault-tolerant manufacturing architecture.

through fair, nonblocking merging and distribution junctions, and transferring structures back to belts again will be outline and shows how fail-stop units can be joined by redundant tracks and junctions in such a manner that any single unit, track segment, or junction can fail without interrupting the transformation of inputs into outputs. In particular, the system can continue to work without error so long as one unit in any set remains operational. (The presence of redundant units decreases the operational frequency of each, decreasing energy dissipation per operation.)

In the typical case, failure rates are dominated by failures in the mechanosynthetic units rather than in the transportation network. In this case, the probability P_{syst} that a system remains operational is related to the probability P_{unit} that a unit remains operational by the expression

$$P_{syst} = \left[1 - (1 - P_{unit})^{N_{unit}}\right]^{N_{set}} \quad \text{...(1)}$$

where N_{unit} and N_{set} are the number of units in a set and sets in a system, respectively. (Unit failures are assumed to be independent; substantial spatial separations can be provided.) A more readily evaluated expression is

$$P_{syst} \approx \exp\left[-N_{set}\left(1-P_{unit}\right)^{N_{unit}}\right] \text{ where } P_{unit} \approx 1. \qquad ...(2)$$

Manufacturing-system Architecture

This section outlines an architecture for a system capable of manufacturing macroscopic objects. The subsystem capacities are chosen to permit the conversion of a feedstock solution consisting of small organic molecules into ~1 kg product objects of ~0.2 m dimensions in a cycle time of ~1 hour. The flow of materials proceeds through molecule sorting and orientation, preparation of reagent moieties, several stages of convergent assembly using mill-style mechanisms and several stages of convergent assembly using manipulator-style mechanisms.

This section focuses on mechanosynthesis and assembly, omitting the details of supporting systems. For example, although they would be essential in a detailed design, the specifics of a ~1 kW cooling system, a ~0.001 kg/s feedstock solution supply system, transportation and sealing mechanisms for product delivery, and so forth are peripheral to the central issues of molecular manufacturing. The convergent assembly processes can be performed without undue geometrical problems in material flow; accordingly, a reasonable estimate of overall system volume can be had by summing the volumes of the assembly workspaces without describing a particular three-dimensional layout. Maintenance of vacuum integrity has been discussed. Compressive structures made of diamondlike materials can support terrestrial atmospheric pressures with masses of «1 kg/m³, and can be designed in many ways. Power conversion between mechanical and electrical forms can be performed with high efficiency and (for reasonable power levels) with negligible mass and volume, imposing few constraints in this context.

The key issues in a manufacturing architecture of this sort center around the assembly mechanisms themselves: their numbers, kinds, reliabilities, operating frequencies, masses, volumes, and so forth.

Products, Building Blocks, and Assembly Sequences

Using fixed, mill-style subsystems throughout a manufacturing system, from the moietal level to finished products, would limit it to making products of a single kind. Using flexible, manipulator-style subsystems throughout would permit independent control of every detail of a product, permitting an indefinitely large range of structures at all scales, but at the cost of relatively slow, inefficient manufacturing processes. Neither of these extremes is likely to prove desirable.

Building blocks and product diversity

Typical macroscopic products can conveniently be made from modular building blocks, many identical to one another. Micron-scale building blocks are small enough to make almost any macroscopic shape in ordinary use today within better tolerances than those provided by conventional machining. From sets of building blocks that can be assembled to make solids of differing strength, stiffness, conductivity, and so forth, objects can be made that exhibit material properties that equal or better those of present industrial products. Systems that take more direct advantage of nanoscale structures can still use extensively duplicated building blocks.

Products containing a macroscopic quantity of computational hardware will typically contain many identical CPUs, memory arrays, and so forth. A product containing a macroscopic quantity of mechanical structure will typically contain many identical regions of structural material. Motors, gearboxes, vacuum pumps, data-bus segments, photovoltaic cells, storage batteries, conductors, struts, connectors, manipulator arms, reagent-processing subsystems-all can be constructed on a micron or submicron scale, and all are candidates for extensive duplication in macroscopic systems.

Accordingly, a wide range of macroscopic products can be made with good efficiency by using mills to make a diverse set of building blocks in the 10^{-7} to 10^{-6} m size range and then using manipulators to assemble them into macroscopic products. Inclusion of a secondary production capability that applies manipulators to smaller building blocks (at the moietal level where necessary), can provide unusual building blocks without requiring a dedicated mill mechanism. The exemplar system makes the simplifying assumption of a hand-off from mills to manipulators at a single block size, 10^{-6} m.

Alternative assembly sequences

The quantitative summary of the exemplar architecture assumes a strictly convergent assembly sequence in the mill mechanism: each moiety and each subassembly is destined for a unique product at the 10^{-6} m block level. The reagent preparation mill system, in contrast, produces standard moieties that can be used by any of a wide range of assembly mills. There is good reason to extend this pattern to the subassembly level, thereby permitting flexible allocation of the output of subassembly mill mechanisms to any of several higher-level assembly mills. Doing so enables a system to produce a greater range of 10^{-6} m scale blocks with a lesser increase in total mill system size.

A further assumption in the quantitative summary is that each assembly unit combines blocks of similar sizes to make a much larger structure. There is no reason to treat this as a constraint, and there are substantial advantages to assembling components of widely differing sizes. For example, after assembling two large blocks with an adhesive interface, moietal-scale modification of the edge of the interface region might be necessary to prepare a smooth surface spanning the two blocks. Macroscopic assembly processes frequently add small components to large structures.

Adjustable blocks

A large number of blocks will serve simple structural functions, holding active components such as motors, gearboxes, sensors, and the like in fixed geometrical relationships to one another. This requires struts of diverse lengths and joints of diverse angles. There is, however, no reason why structural elements of each needed length and angle must be manufactured by a distinct assembly mill, starting at the moietallevel.

It is straightforward to build struts and joints that contain internal sliding interfaces permitting them to be extended or twisted to assume a wide range of lengths or angles, and then twisted or compressed (respectively) to bring internal adhesive interfaces into contact, thus producing a stiff structural member with a particular geometry. In this manner, a mill system can make identical, adjustable members that each of several subsequent mills (or manipulators) can transform into one of many fixed, specialized structures.

Throughput, Delays, and Internal Inventories

If all components operated at their stated frequencies, handling components with a density of 1000 kg/m^3, system throughput would be 0.001 kg/s. The five mill stages operate at high frequencies, performing relatively small numbers of operations per output object; the maximum delay for material passing through these five stages is $\sim$0.1 s, implying a $\sim 10^{-4}$ kg inventory of materials in process.

The three manipulator stages operate at lower frequencies, performing relatively large numbers of operations per output object; each requires 1000 s to complete its operations, and contains a time-average product mass of 0.5 kg. Allowing for feasible overlap in the operating times of the manipulator stages, the delay between process initiation and delivery of a 1 kg product object, starting from a feedstock solution, can be somewhat less than one hour.

Mass and Volume

The volume of the first stages—input ordering, reagent preparation, and mill mechanisms—can be estimated by applying an estimated mean density to the tabulated masses. Using 100 kg/m^3 yields a volume of $\sim 1.8 \times 10^{-4}$ m^3. (System flexibility could be increased by devoting ~ 10 times this mass and volume to mill mechanisms, most idle in a typical production process; other system parameters would remain essentially unchanged.) The volume of the manipulator stages depends on the size of the product structures, if these are to be totally enclosed in the workspace rather than constructed and extruded incrementally. A wide range of 1 kg products should fit (perhaps in a folded or partially disassembled configuration) within a 0.01 m^3 workspace. If the solid parts of a 1 kg product object are as dense as diamond, then these parts occupy ~ 0.03 of the workspace volume. Assuming that workspace volumes are nonoverlapping, but can be used for conveyance of parts from other workspace volumes, then the total volume required for the three manipulator stages is ~ 0.03 m^3. Multiplying by 1.66 to allow for packing inefficiencies, cooling channels, and so forth yields a total estimated volume of ~ 0.05 m^3.

The total mass of the manufacturing system components is <0.12 kg. Allowing 1 kg/m^3 for internal compressive structure to support external atmospheric pressure adds ~ 0.05 kg; allowing 0.1 kg/m^2 to provide for a sturdy case, rubber feet, and the like adds ~ 0.08 kg. These masses sum to <0.25 kg. Allowing as much as 0.75 kg for other subsystems yields an estimated total system mass <1 kg.

System Lifetime

Radiation damage chiefly threatens the small yet relatively specialized devices in the first three stages of the mill system. Earlier stages are massively redundant; later stages use components large enough to permit radiation-insensitive designs. Applying the previous radiation-damage assumptions, units with a radiation-sensitive mass of $\leq 2 \times 10^{-18}$ kg have $P_{unit} = .99$ after ten years. The radiation-sensitive portions of mill stages 1-3 have a lower unit mass, with $N_{unit} = 10$, $N_{set} \approx 10^{15}$, and hence $P_{syst} \approx 1$ after ten years. Increasing N_{unit} to 20 raises the expected lifetime to over a century. All radiation-sensitive units are assumed to be fail-stop and linked by redundant transport networks. An enclosure of this size can easily be made UV opaque.

Feedstock Materials

A specific manufacturing mechanism will accept raw materials consisting of a specific set of small molecules in solution. Typical products require large quantities of C, moderate quantities of H, N, O, F, Si, P, S, and Cl, and lesser quantities of several other elements. These could be provided by a single solution, or by different solutions supplied through several ports. In designing a system, the choice of input compounds is largely a matter of cost and convenience.

By-products

Extended diamondlike structures have a smaller mass fraction of hydrogen than do typical small organic molecules: small molecules have a larger surface-to-volume ratio, and satisfaction of surface valences is most commonly achieved by inclusion of monovalent atoms, chiefly hydrogen. Fullerenes, cyanogen, perfluoroalkanes, and other unusual molecules are among the exceptions to this pattern. Aside from this, there is no strong reason why the elemental composition of the inputs should differ from that of the product, and hence no strong reason why there should be substantial by-products from a molecular manufacturing system. Excess hydrogen could be delivered as water, in a process that consumes atmospheric oxygen and (presumably) produces mechanical energy as a by-product.

If a mechanism is supplied with a single solution of fixed composition, and must from this produce a variable mix of products, the (nonhydrogen) elemental composition of the products will in general fail to match that of the input solution, and the less-consumed compounds will accumulate. The resulting solutions can be recycled by adding the compounds in which they have become deficient.

Energy Output and Dissipation

The energy dissipated by mills is dominated by moietal operations as 1.5×10^6 J/kg; a further allowance of 5×10^5 J/kg is made for energy dissipated during block assembly. The mechanical aspects of manipulator operations result in comparatively negligible energy dissipation, as do the operations in molecular sorting and ordering (save for entropic effects included in the overall thermodynamic calculations that follow).

The energy dissipation resulting from computation used to direct manipulator operations is hard to estimate accurately, but can be made quite small in the exemplar architecture. The control of motions can be relatively direct: with compiled instructions, no run-time planning is necessary, hence the computations can be as simple as loading appropriate internal-coordinate settings into registers in controllers. A relatively computation-intensive process (perhaps analogous to the interpretation of a page-description language) might execute $\sim 10^6$

instructions per block placed. An estimated energy dissipation of $\sim 10^{-16}$ J per instruction, and a 1 kg object contains $\sim 10^{15}$ cubic-micron blocks; hence executing even 10^6 instructions per block would increase the total energy dissipated by only 10^5 J/kg.

Transformation of hydrogen-rich organic feedstock molecules and oxygen into typical diamondoid products and water is an exoergic, entropy-reducing process. For example, a process with acetone as the primary carbon source and diamond as the chief product liberates a total energy equaling $\sim 1.7 \times 10^7$ J/kg of useful output, while decreasing local entropy by $\sim 5.7 \times 10^3$ J/kg·K (approximating the entropy of the product object as zero). The change in free energy accordingly is $\sim 1.5 \times 10^7$. Allowing for $\sim 3.1 \times 10^6$ J/kg of free energy dissipated, the system produces $\sim 1.2 \times 10^7$ J/kg of surplus energy in the form of mechanical work. This could be conveniently disposed of by delivery to an external power distribution system in the form of electrical energy.

Waste heat from the process results from both inefficiencies and computation ($\sim 3.1 \times 10^6$ J/kg) and from the reduced entropy of the products relative to the feedstocks ($\sim 1.7 \times 10^6$ J/kg). Thus, a system producing 1 kg/hr of diamondoid products dissipates an estimated 1.3 kW of waste heat. The required cooling capacity can be provided by fan-driven air flowing at a rate of 0.1 m^3/s with $\Delta T = < 15$ K between the intake and exhaust ports. Cooling channels can be placed in close proximity to the chief heat sources, permitting good thermal contact.

Information Requirements

A typical product of the exemplar architecture consists of $\sim 10^{15}$ building blocks of $\sim 10^{-6}$ m linear dimensions. To specify a product requires specifying the type and position of each block. If there are $< 10^9$ different kinds of block, then specifying the type of a block requires < 32 bits of information (substantially less, with efficient encoding). The total quantity of information required per block need seldom be much larger than this.

An analogy can be drawn to a pattern of type on a page: letter forms are complex, but standardized; with a little additional information (e.g., column widths, line spacings, superscript and subscript codes) a sequence of 8-bit codes specifying a sequence of letters determines the two-dimensional arrangement of those letters on a page. This may, however, depend on large auxiliary tables of information on letter widths, spacings for kerned pairs, and so forth. In a similar manner, a sequence of codes specifying blocks can be combined with auxiliary information describing the acceptable geometrical relationships between adjacent blocks to provide most of the information required to determine the three-dimensional arrangement of those blocks in space. This approach exploits stereotyped relationships among adjacent blocks to generate precise positioning information through simple computations.

Specification of an object should thus require substantially fewer than 100 bits per block, or $<10^{17}$ bits per product object. Since molecular tape storage systems can hold $\sim 10^{28}$ bits/m^3, a volume of $<10^{-5}$ m^3, containing <0.01 kg, can contain sufficient information to specify the structures of 10^6 product objects. (Note that the presence of repeated substructures above the block level, either within or between objects, can greatly reduce information requirements.)

Manufacture of Manufacturing Systems

Since most of the internal volume of the exemplar architecture is devoted to open workspaces for manipulators, it should be feasible to design a system that can be folded from linear dimensions of <0.4 m to linear dimensions of ~ 0.2 m, or (more to the point) that can be unfolded from the smaller dimensions to the larger. How a continuous, gas-tight wall can be expanded by a greater factor? This places the exemplar architecture within both the size and mass range of its own products.

With the use of programmable manipulators to build a diverse set of structures from a smaller set of building blocks, the output of a set of specialized mills can be used to build an identical set of mills, as well as many other structures. Accordingly, a suitably designed and programmed molecular manufacturing system, built along the lines of the exemplar architecture, can be used to build objects that, when unfolded, are substantially identical to itself.

The design of machines able to construct copies of themselves was described in von Neumann and Burks (1966), and many variations on this theme have been reviewed in Freitas and Gilbreath (1982) and the included references. It may seem somehow paradoxical that a machine can contain all the instructions needed to make a copy of itself, including those selfsame complex instructions, but this is easily resolved. In the simplest approach (as implicitly adopted here), the machine reads the instructions twice: first as commands to be obeyed, and then as data to be copied. Adding more data does not increase the complexity of the data-copying process, hence the set of instructions can be made as complex as is necessary to specify the rest of the system. By the same token, the instructions transmitted in a replication

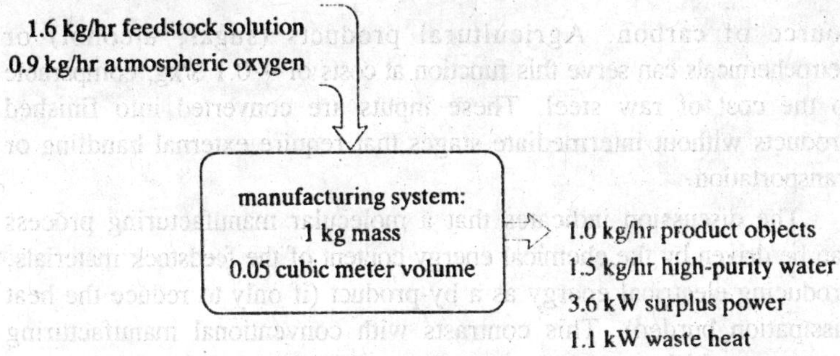

Fig. 6.6. The inputs and outputs of the examplar manufacturing system.

cycle can specify the construction of an indefinitely large number of other artifacts.

Comparison to Conventional Manufacturing

The characteristics of molecular manufacturing systems are best understood in comparison to the characteristics of conventional manufacturing systems. The following comparisons embrace required inputs of materials and energy, byproducts, sizes of internal components, frequencies of internal operations, overall productivity, characteristics of the resulting products, and costs of production. Differences in the technology base required for implementation are profound, which describe strategies for developing molecular manufacturing capabilities.

The discussion in this section is not a basis for further analysis of molecular machinery, manufacturing, or computation; rather, it attempts to evaluate some of the consequences of the previous analysis. In characterizing such broad topics as the entire range of product characteristics and the entire range of potential by-products, the descriptions are based on informed estimates. Most other conclusions are direct consequences of the previous analysis, supplemented by commonplace knowledge of present technological capabilities.

Feedstocks and Energy Requirements

Defining what one means by a "feedstock" for conventional manufacturing can be difficult. A factory making products for household use may consume inputs as simple as sheet metal and wire, or as complex as integrated circuits and prefabricated assemblies. A system of factories considered as a whole typically includes smelters, steel mills, and the like, starting with crude feedstocks (iron ore, coal).

In molecular manufacturing, a mixture of simple compounds can serve as a feedstock. Quantitatively, the largest requirement is for a source of carbon. Agricultural products (sugar, alcohol) or petrochemicals can serve this function at costs of ~0.1 \$/kg, comparable to the cost of raw steel. These inputs are converted into finished products without intermediate stages that require external handling or transportation.

The discussion indicates that a molecular manufacturing process can be driven by the chemical energy content of the feedstock materials, producing electrical energy as a by-product (if only to reduce the heat dissipation burden). This contrasts with conventional manufacturing processes, which consume energy not contained in their inputs. Further, both the energy and material input requirements per unit of product functional capacity are substantially lower than those for conventional manufacturing processes, owing to greater functional capacity per unit product mass.

By-products and Recycling

From the discussion, it should be clear that molecular sorting and processing mechanisms can perform transformations like those now performed in chemical plants, but with greater efficiency and control. The preparation of feedstock solutions need not generate by-products in uncontrolled forms; indeed, systems can be designed to produce no by-products at all, aside from some set of compounds containing any unwanted elements present in a crude input material. A general strategy for accomplishing this would be to extract and process small, soluble molecules from a mixture, then use conventional, nonspecific techniques (heat, oxidation, acids) to break down any residue into small, soluble molecules. Design studies to date suggest that relatively exotic and toxic materials (lead, mercury, cadmium) need play no role in molecular manufacturing processes or products.

Many products can be built as aggregates of modules joined by reversible means. These can presumably be recycled with little energy cost by disassembly and reassembly operating at the scale of micron-scale blocks. Failing this, almost any product can be recycled into small molecules subjecting it to chemical attack in a closed vessel (e.g., by oxidation in a sealed incinerator, followed by other processes to break down any residual solids). These small molecules can then be sorted and reprocessed.

Internal Component Sizes and Frequencies

In conventional manufacturing, human hands and other devices on a 0.01 to 1 m scale perform operations at frequencies that typically range from 10 to 0.1 Hz. In molecular manufacturing, the most numerous operations are performed by devices on a 10^{-9} to 10^{-7} m scale, operating at frequencies of $\sim 10^6$ Hz. In making macroscopic objects by convergent assembly, a modest number of operations are performed at frequencies of ~ 10 Hz.

Productivity

The physical productivity of a system can be measured in various ways. One is the time required for a system to produce outputs with a total mass equaling its own (by this standard, a short piece of pipe can be enormously productive). Another basis, less precise but still significant, is the length of time required for a system to produce outputs with a total complexity equaling its own, starting with simple inputs. If this criterion is applied to the materials processing and manufacturing sector of an industrial economy, the result corresponds to the minimum doubling time for the total capital stock, and is presumably on the order of a year or more, $\sim 10^8$ s. For the exemplar molecular manufacturing system, the equivalent time is $\sim 10^3$ s. This disparity stems from the difference in the characteristic frequency of operations.

Some Feasible Product Characteristics

Precision and reliability

Most products made by present manufacturing systems are poorly ordered: they have either amorphous microstructures or crystalline microstructures that contain a high density of atomic defects, impurities, dislocations, grain boundaries, and microcracks. Semiconductor substrates come closest to perfection, but even so, most devices are formed by imprecise processes such as diffusion of impurities, nonepitaxial deposition processes, etching, and the like. In addition to having poorly controlled microstructures, macroscale objects have shapes that are rarely precise to $<10^{-6}$ m. Without robust designs and careful process control, statistical variations in defect size can make products unreliable. Designs that tolerate material defects usually must sacrifice performance to do so.

Molecular manufacturing systems can make well-ordered products, save for scattered atomic-scale defects caused chiefly by background radiation. The materials-, component-, and system-level analyses have shown how this more precise control can yield systems of high performance. Designs that tolerate the residual flaws caused by radiation damage commonly sacrifice performance, but the lesser size and density of the flaws results in lesser sacrifices. For structures small enough to avoid damage, or large enough to be inherently tolerant of it, the reproducibility of properties and performance can approach perfection. This improves reliability.

Strength and stiffness

It is well known that intrinsically strong materials (e.g., ceramics such as aluminum oxide, silicon carbide, and diamond) are brittle, and that brittle materials placed in tension are sensitive to microcracks. Accordingly, the ability of current engineering practice to exploit the high intrinsic strength of these materials is sharply limited by the inability of present manufacturing processes to make them without defects. The ability of molecular manufacturing to build structures with only atomic-scale defects permits these materials to be used at nearly their full theoretical strength. Diamond, in particular, has ~55 and 75 times the ratio of tensile strength to density of strong steel and aluminum alloys, respectively (assuming a conservative 50 GPa tensile strength for diamond, and tabulated values for AISI No. 9255 steel, hardest temper, and 7178-T6 aluminum).

The design of minimum-mass compressive structures often is constrained by stiffness, and the ratio of Young's modulus to density for diamond exceeds those for steel and aluminum by factors of 12 and 15, respectively. The design of compressive structures is often further constrained by buckling, and strategies for preventing buckling are often constrained by the scale of initial imperfections and by the difficulty of fabricating the optimum shapes (which can be intricate, and can pose minimum-gauge problems). Finally, it is worth noting that small moving parts, high power-density actuators, and fast sensing and control systems can be combined to make macroscopic objects that actively resist deformation, thereby simulating an extremely high modulus material within certain bounds of frequency response, length scale, and applied load.

Other material properties

Conventional manufacturing processes are constrained to work with materials that result from bulk (or biological) processes. Although it is difficult to characterize the limits of what such processes can make (particularly through organic synthesis), they control comparatively few variables: chiefly the time history of bulk composition, temperature, and pressure, augmented by shear rates, magnetic fields, and electric fields. Molecular manufacturing processes, in contrast, control a comparable number of variables *per moiety* during the fabrication process. This strongly suggests that the range of feasible materials (and therefore material properties) is broader. In some instances, however, bulk processes can presumably make materials that are at or near the limits of performance in particular respects.

The processes and exemplar architecture described in this chapter assume a high vacuum environment, permitting free use of highly active reagents with no special provision for local enclosures. This limits the output to product structures of high stability, and in particular, of negligible vapour pressure. Most high-strength materials and many polymers and elastomers meet these criteria, but many familiar products do not. However, describe the use of mechanosynthetic processes in solution environments (with suitable constraints on reagents, mechanisms, and so forth). High vacuum is convenient and often desirable, but it should not be taken as a necessary condition for all molecular manufacturing processes; indeed, thorough control of the immediate environment of a reaction site can be provided in a liquid medium, as it is in enzymes. Convergent assembly via blocks of intermediate scale can impose constraints on structure, but the feasibility of joining a pair of blocks to form dense, continuous diamondoid structures suggests that these constraints are not severe, in practical terms.

Size

Manufacturing processes today rarely produce features less than 100 nm in size, and most products are macroscopic, having features on a scale of 0.001 m or more. Molecular manufacturing can construct objects with ~ 0.3 nm features, placing $\sim 10^8$ distinct atomic features in a volume of $(100 \text{ nm})^3$. At the opposite end of the size spectrum, both conventional and molecular manufacturing processes built along the lines of the exemplar system can construct objects of indefinitely large size (in both instances, by assembling macroscopic components to form larger structures).

Energy conversion

A class of submicron electrostatic motors capable of converting between mechanical and electrical power in either direction at a power density of $\sim 10^{15}$ W/m^3 and an estimated efficiency $>.99$. This power density is several orders of magnitude greater than that of electric motors produced by current manufacturing processes.

It uses previous calculations regarding mechanosynthetic processes to conclude that chemical and mechanical energy can be interconverted with an efficiency $>.99$ at a power density of $\sim 2 \times 10^9$ W/m^3, or $\sim 8 \times 10^6$ W/kg. If suitably low-dissipation designs can be found (this chiefly entails avoiding major nonisothermal compression losses), then higher speeds of motion (e.g., ~1 to 10 m/s, rather than ~0.004 m/s) will be readily achievable with good efficiency, and the power density values that are achievable will increase to ~10^{12} W/m^3, or ~10^{10} W/kg.

Conversion of optical to electrical energy has applications in power supply. Molecular manufacturing techniques can be used to make multilayer, multiband-gap photovoltaic cells; structures of this sort can convert solar energy to electrical energy with efficiencies substantially greater than .3.

Conversion of electrical to optical energy has applications in display and illumination. The use of quantum-well nanostructures is already of technological interest for improving the performance of light emitting diodes and solidstate lasers, hence the nanoscale structural control provided by molecular manufacturing techniques should lead to significant improvements. Isotropic light-emitting structures have no minimum scale, but owing to diffraction, good directionality requires a size that is large compared to an optical wavelength.

Energy storage

Nearly reversible mechanochemical energy conversion can serve as a basis for energy storage. For example, a device that nearly reversibly transforms butane and oxygen into water and carbon dioxide can store ~4 × 10^7 J/kg, based on the mass of the initial butane, or ~9 × 10^6 J/kg, based on the total mass of the reactants. Rates of energy storage and release are (note that the power conversion rates possible in macroscopic systems will commonly be limited by cooling). Systems of this sort have substantially higher energy densities than those of storage cells that can be produced using present manufacturing technologies, and power densities that are orders of magnitude higher.

Computational capacity

Molecular manufacturing processes can be used to construct computational systems (or better, given the conservatism of the design rules employed there). In terms of component densities, some of the relevant system-level parameters include ~10^{19} CPUs/m^3, ~10^{25} bits/m^3 (for RAM storage), and ~10^{28} bits/m^3 (for tape storage). Individual CPU speeds can equal or exceed 10^9 instructions per second with an energy consumption of ~10^{-16} J per instruction. Aside from individual CPU speed, each of these parameters exceeds by several orders of magnitude the performance possible in systems made by conventional manufacturing processes.

Manufacturing Costs

Feasibility of making cost estimates

Estimating the cost of a conventional manufacturing process is difficult unless the product closely resembles one for which experience exists. Product costs depend on the costs of components which mayor may not be in production. The costs of establishing, staffing, and managing a production process–including the controls necessary to ensure adequate product quality—can be hard to estimate. In the aerospace industry, in particular, the costs of testing and documentation for new technologies are huge. Where a technology differs greatly from existing practice, estimates of manufacturing costs have traditionally been little more than guesses.

The relationship between complexity and manufacturing costs highlights the unusual nature of cost estimation in molecular manufacturing processes. In conventional manufacturing processes, costs increase with the complexity of the product being made: more intricate systems require more parts and manufacturing operations. In molecular manufacturing processes, each moiety is treated as a distinct part, regardless of the apparent complexity or simplicity of the product. Accordingly, production cost will be essentially independent of complexity.

In molecular manufacturing, many of the traditional costs of building complex systems are avoided. Most of the costs that remain are sufficiently well defined that one can place upper bounds on them. These costs are estimated on a per-kilogram basis; for comparison, most manufactured products today fall in the cost range between ~ 10 \$/kg (e.g., automobiles, appliances) and ~ 10^4 \$/kg (e.g., aerospace vehicles).

Materials

As has been discussed, the chief anticipated feedstock materials for molecular manufacturing (C, N, 0, H) are available in bulk compounds for costs of ~ 0.1 \$/kg (at a substantial energy cost, they are available from atmospheric gases). Others that may play substantial roles (Si, P, S, F, Cl) are also available in reasonably inexpensive forms. Precious metals (e.g., of the platinum group) can be expected to be applied as catalysts in mechanochemical processes, but if used at only 10^4 cycles per second, a kilogram of platinum atoms (costing ~ 10^5 \$) can be used to process > 10^{10} kg of material per year, adding little to the product cost.

Energy

Using typical organic feedstocks, and assuming oxidation of surplus hydrogen, reasonably efficient molecular manufacturing processes are net energy producers. At a typical price for electrical energy today, ~0.1 \$/kW·hr, the value of the by-product electrical energy would usually exceed the cost of the feedstock materials. The value of by-product energy will be ignored in the present cost estimate.

Waste disposal

The only wastes from the manufacturing process itself are high-purity water and a modest amount of heat rejected to the atmosphere (for a 1 kg/hr system, this is roughly equal to the heat produced by sunlight striking a square meter of dark surface). Neither of these would now be regarded as a significant form of waste. The cost of waste disposal from the preparation of feedstock materials is (within present environmental standards) reflected in the cost of those materials.

Labour and distribution

Between the input of raw materials and the removal of finished products, no labour is necessary. Both of these steps can be regarded as distribution costs falling outside the scope of manufacturing proper.

Land

The exemplar manufacturing system produces its own mass in product in less than an hour. Conventional manufacturing systems (which must in a fair comparison be taken to include factories that supply parts) take orders of magnitude longer to produce their own mass in product. Assuming comparable ratios of plant and product densities, this indicates that molecular manufacturing systems require orders of magnitude less space (and land) per unit of productive capacity. Costs of land will be reduced by the same factor.

Capital

In current manufacturing systems, the cost of capital goods is chiefly a manufacturing cost. In considering a new manufacturing system, this introduces a circular dependency in cost estimation: the cost of producing capital goods depends on the costs of materials, energy, waste disposal, labour, distribution, and capital goods. The equilibrium contribution of the cost of capital goods to the cost of later capital goods can be estimated as follows: Let r be the real interest rate (expressed as a fractional rate of return per year), c be the noncapital cost of producing a unit of capital goods, and t be the time (in years) for a unit of capital goods to produce another unit of capital goods (here treated as a batch process). The total cost C of a unit of capital goods is then

$$C = c + C[\exp(rt) - 1] \quad \text{...(3)}$$

$$\approx \frac{c}{1 - rt}, \quad rt << 1 \quad \text{...(4)}$$

In conventional manufacturing, $r \approx 0.1 \text{ yr}^{-1}$, and $t \approx 1$ yr (or more, particularly if system boundaries are drawn to include factories that supply parts), and the contribution of the cost of capital goods is substantial. In molecular manufacturing, $t \approx 10^{-4}$ yr, and these costs are negligible. In this connection, it might be noted that the cost of capital goods contributes to the cost of materials and energy, but the cost benefits of changes in the resulting circular dependencies will not be considered here.

Total costs of manufacturing

The preceding discussion indicates that the basic cost of production, here taken to exclude the costs of development and distribution (and of such imponderables as taxation, licensing agreements, and insurance), will be almost wholly determined by the cost of materials. The relevant materials presently cost ~0.1 to 0.5 \$/kg; this range can thus be taken as an upper bound on the basic cost of producing objects using a mature molecular manufacturing technology base. These costs are low enough for molecular manufacturing to be competitive in making a wide range of products.

Design and Complexity

The design of systems that organize complex patterns of activity among reliable, submicron devices operating at millions of cycles per second is an everyday activity in computer science. Although molecular machines and manufacturing differ greatly from microelectronics and computation, many similar issues arise. It is reasonable to expect that computer scientists will play a leading role in the development of molecular manufacturing systems at levels above device design. Accordingly, this section repeatedly compares established concepts in computer science with the anticipated needs of molecular manufacturing.

Device design will likewise depend on computer science. Although experimentation is likely to playa large role, simulations and computer-aided design (CAD) tools will be essential. Complex systems will become possible only when an adequate library of simple components and operations has been accumulated. The following discussion will assume the previous development of such a library, thereby sharpening the focus on issues arising from complexity.

Part Counts and Automation in Design and Computation

Many of the numbers that describe the manufacture of macroscopic objects by molecular means are enormous by most present standards. Building a cubic-micron block requires $\sim 10^{11}$ moietal assembly operations. Building a kilogram object from cubic-micron blocks requires $\sim 10^{15}$ block assembly operations. A team of a thousand people each making one design decision per second would require a century of 8-hour days to plan a single cubic micron, to say nothing of a kilogram: in a practical design process, most of these operations must be specified without direct human attention. Most operations must be as routine as painting a pixel in a raster output device or executing an instruction in a piece of code.

For comparison, typesetting this volume required that a complex pattern of $> 10^{10}$ pixels be painted black or white; these operations were directed by a set of single-purpose PostScript programs that were automatically generated from $> 10^{8}$ bits of data describing text and graphics. The software used to prepare this data executed $> 10^{13}$ instructions (though mostly while waiting for inputs). The software itself consisted of $> 10^{7}$ instructions, written by several teams to form separate programs that nonetheless cooperated to produce a single product. It is reasonable to assume that molecular manufacturing, which likewise generates complex patterns through repetitive operations, will involve broadly similar operations controlled by broadly similar software systems.

Design of Components and Small Systems

Today, most machine code that directs computer operations is generated automatically from a more abstract description in a high-level language. The compiler that does this works by piecing together small, standardized fragments of machine code (consisting of one or more instructions) to implement the desired operations. These fragments and the rules for generating sequences of them are the result of human attention, but the results of this labour are used repeatedly without human attention. In a similar fashion, component-level nanomechanical designs and rules for their composition can be designed with human attention, then later combined by a *hardware compiler* to implement desired functions.

Examples of direct atom-by-atom design of molecular mechanical devices are scattered through this volume. Some designs, however, show the early results of automation, in which nanometer-scale components are described by languages that specify regular arrays of atoms and bonds. A good example is R. Merkle's prototype computer-aided design (CAD) software that starts with a single typed line of input parameters and generates the atomic coordinates of a regular, three-dimensional diamondoid structure that can contain many thousands of atoms and bonds.

The design of less regular structures can also be automated. A precompiled library of Kaehler brackets can be used to choose bridging structures between surfaces. Simulated annealing methods (with change of valence and atom type as an operator) hold promise for generating regions of amorphous structure that meet particular boundary conditions.

Eventually, molecular CAD tools will combine principles like those applied in Merkle's and Kaehler's prototypes with features like those found in commercial CAD systems for macroscopic design, such as support for continuum descriptions of objects containing many atoms. With these CAD tools (and faster machines, better molecular potentials, and feedback from physical experiments) it will be possible to design components and systems of the complexity that can be managed today in the macroscopic world. These will include manipulator arms, computers, and molecular mill systems using multiple belts, rollers, and cam surfaces. Components and systems of this complexity can then serve as building blocks in the design of more complex systems.

Automated Generation of Synthesis and Assembly Procedures

Retrosynthetic analysis of chemical syntheses

Once the structure of a component is specified, a synthesis must be found or the component must be redesigned. Software for the automated generation of candidate chemical syntheses has been in use since the 1960s. These programs apply *retrosynthetic analysis*, the development and application of which won E. J. Corey the 1990 Nobel prize for chemistry. The central idea of retrosynthetic analysis to plan the assembly of a complex molecular structure by considering how it could be *disassembled* into small, available molecules using imaginary operations, each the *reverse* of a feasible synthetic step. The resulting disassembly process is therefore the reverse of a feasible synthesis.

Mechanosynthetic operations are equally subject to retrosynthetic analysis (but largely free of the complication of competing reactions). Each synthetic step can be regarded as the reverse of an etching reaction, and the synthesis designed (in reverse) as the etching away of the desired structure. Considering the mechanosynthesis of diamondoid structures, Soreff (1992) has suggested that a study of actual etching processes (e.g., of diamond by fluorine) can indicate geometries and transition states for synthetic steps; the forward-direction reagents would be analogous to the hypothetical etching products, but with bond strengths (etc.) favouring deposition over etching.

Hierarchical decomposition of larger structures

If an object is to be made from an array of parts, it (and its assembly procedure) can be designed by dividing it into regions, each made of subregions, and so on down to the level of the individual parts. The divisions chosen during the design phase must create interfaces that permit assembly during the manufacturing phase. If the final product were fully specified at the outset, this hierarchical decomposition approach would be a form of retrosynthetic analysis.

In conventional engineering, however, the specification of the final product is influenced by the nature of the available parts and assembly operations. Abstractly, this means that the hierarchical decomposition is developed together with the design, rather than afterward. In dividing a system into regions, the interfaces can be altered as the design becomes more detailed. A design process that includes an assembly-oriented hierarchical decomposition can directly generate the outlines of an assembly plan, and can provide a framework for dividing the computational load of the design process over a large number of processors. (For other purposes, designing in terms of functional rather than geometrical decompositions will be important, requiring multiple representations of the design.)

Shape Description Languages and Part Arrays

A *shape description language* can describe a shape so as to enable an interpreter driving an output device to make a corresponding physical structure. In a program in the PostScript language, which describes two-dimensional shapes, a set of instructions can describe a circle by its radius, colour, and border. A language for describing three-dimensional shapes might similarly describe a sphere by its radius, material, and surface structure. With PostScript driving a printer, the circle is approximated by printing an array of dots, fixed with respect to a page; with a three-dimensional language driving a molecular manufacturing system, a sphere would be approximated by assembling an array of blocks, and, unless specified as joined to another object, would remain mobile. In either case, the complexity of the description is independent of the size of the object and hence of the number of pieces from which it is made.

Three-dimensional shape description languages are standards in computer-aided design systems for mechanical engineering. An interpreter operating on a CAD description could generate instructions for a molecular manufacturing system, directing it to make a corresponding set of solid objects. Using cubic-micron blocks that can be composed to make solid regions of widely varying properties, the output structures could form a wide variety of high-performance systems with submicron control of component shapes. Not all CAD descriptions will correspond to feasible assembly sequences, but many structures can be made to fall within the required design rules (for example, that each object be supported while it is being built, and that each object be subject to a hierarchical decomposition forming suitable interfaces).

Compilers

Assembly-process compilers

Given a fully specified design that meets the constraints necessary to permit assembly, there remains the task of specifying the operations of the assembly process. If the product is to be made from micron-scale blocks, this amounts to specifying the transport of blocks of the correct types to the correct assembly workspaces in the correct order, and specifying the sequence of motions to be executed by manipulators in putting the blocks together. The design of structures and assembly procedures by hierarchical decomposition directly generates a tree of assembly steps. If this tree has been chosen to generate parts of the appropriate sizes and numbers, then it can be mapped onto a manufacturing layout. The vast number of manipulator motions to be specified at the finest, earliest, and most-dispersed levels of the assembly process can be planned in parallel by identical software systems running almost independently on separate processors. The result of this parallel assembly-process compilation is a set of instructions that, when executed by the transportation system and manipulator controllers of a manufacturing mechanism, will result in the assembly of an object corresponding to the initial design.

Design compilers

A conventional software compiler generates a set of machine instructions from a more abstract specification: a program written in a high-level language. An assembly-process compiler likewise generates a set of machine instructions from an abstract specification, the hierarchically decomposed structure of a particular object (which could itself be specified by a shape-description language). Like a software compiler, it acts as a compiler *with respect to the machine instructions*, although starting with a different kind of specification. A modern *silicon compiler*, in contrast, generates a complex pattern of transistors and conductors from an abstract specification of the properties of a digital circuit. It thus acts as a compiler *with respect to the design of a machine*. The implementation of artifacts as complex as the exemplar architecture seems likely to require the use of analogous design compilers for nanomechanical systems.

There can be no fully general design compilers, capable of converting an arbitrary design specification into a design: not all specifications can be met, and of those that can, not all can be met by the application of compiler-style rules guiding the composition of stereotyped solutions. Like a high-level language compiler or a silicon compiler, a design compiler will accept only specifications for a restricted class of mechanisms in a restricted format.

A compiler for the design of manufacturing systems can only be developed after solutions have been developed to each of the stereotyped, smaller-scale problems that will be encountered. These include:

1. Software modules for performing retrosynthetic analyses of components.
2. Specifications of reaction geometries and environments for effecting standard synthesis steps in mill systems.
3. Designs for subsystems that assemble special components.
4. Designs for structural elements, reagent devices, mill mechanisms, gauges, manipulator arms, computers, communication channels, and so forth.
5. Parameterized descriptions of subsystems for transportation, power distribution, communication networks, and so forth.
6. Design rules for composing subsystems and selecting spatial layouts.

A further requirement, of course, is experimental feedback to permit debugging of system and subsystem designs that fail to work. Note that compiler-aided design need not all be performed by a single tool with seamless end-to-end automation.

The compilation of manufacturing systems will have similarities to the compilation both of digital circuits and of machine code: The product of the compilation process must cause large numbers of operations to be performed in the correct sequence. In designing this product, trade-offs involving time, space, and power must be made, taking into account the relative frequencies and costs of different operations. Preexisting solutions developed for small-scale problems must be described and organized so that they can be applied by algorithms. Throughput, latency, buffers, storage, and transport paths must all be considered. Compact arrangement of parts in space is important (albeit in three dimensions, rather than the two of silicon compilers).

Economics of compiler development

Compilers have not been developed for macroscopic mechanical systems, presumably because either (1) the cost-benefit ratio in developing and using them has been adverse, or (2) they are impossible to develop within the constraints of present computer hardware and software technology. If (2) were true because of software technology constraints, this would cast doubt on the feasibility of developing compilers for nanomechanical systems. The history of software and chip-design compilers sheds light on this situation.

When processors were orders of magnitude slower and memory orders of magnitude more expensive, programs were short, and it was crucial to minimize both size and execution time. Small programs could be described in assembly language, providing instruction-by-instruction control of the machine, and there were great incentives to do so. The first computers became operational in the mid-1940s, but the first high-level language to achieve widespread use was introduced over a decade later, in 1958. Through the 1970s, machine cycles and memory space were scarce enough, and the quality of compiled code poor enough, that assembly language continued to find widespread use in critical programs. Today, faster machines and cheaper memory have enabled the development of programs too large and complex for practical assembly language coding, and improvements in speed, cost, and compilers have combined to make assembly-language coding a shrinking part of software development.

To summarize the abstract lesson: When resources are scarce, systems must be relatively small and simple, and detailed human planning is both feasible and desirable. The incentive to develop compilers is low, and the evolution of compilers good enough to compete with human designers does not even begin. When resources are abundant and inexpensive, detailed human planning cannot organize them on a sufficient scale to exploit their potential. Even inefficient compilers become attractive. They are implemented, and evolve toward greater efficiency.

In macroscopic hardware design today, parts are few and production costs are large. Relatively crude compilers could not compete with human designers in the quality and cost effectiveness of their designs. People can do the job, and do it better. With molecular manufacturing, however, part counts can grow into the trillions and beyond, and per-part costs will fall to minuscule fractions of a dollar. Limiting designs to the number of parts that an unaided team of human designers can manage will be unattractive, if a compiler-aided design team can specify a system a million times more complex and capable. This preference would often remain true even if each compiler-specified system wastes twice as much space, mass, and energy as would a comparable system specified in detail by a competent human designer.

In a design process following the present compiler model, human design will remain dominant at the level of parts and subsystems (in the form of knowledge built into the compiler) and at the level of overall system organization and purpose (in the form of specifications given to the compiler when it is used). The intermediate levels will be designed, with considerable inefficiency, using algorithms and heuristics that represent a workable subset of human knowledge of design principles.

This pattern is, in the early 1990s, already familiar in one hardware domain. Since the early 1960s, the number of components in an integrated circuit has climbed by a factor of $\sim 10^6$, and the manufacturing cost per transistor has fallen to $\sim 10^{-5}$ dollars or less. If a 10^6 transistor design has an expected market of 10^5 units, then every dollar of design cost per transistor adds ten dollars to the price of each chip, yet a dollar cannot buy much time from a human design team. The design cost is often a large fraction of the product cost. In response to this pressure, silicon compiler technology emerged in the 1980s. At first considerably less efficient than a human designer in maximizing speed and minimizing circuit size and cost, silicon compilers nonetheless gained a foothold, then steadily improved, becoming an integral part of the design process. This section has given only a rough sketch of past developments. A more detailed study of the evolution from lower-level CAD systems to silicon compilers could yield further insights into the likely path from today's molecular modeling and mechanical CAD packages toward future compiler-style support for the engineering of nanomechanical systems.

Relative Complexities

Many nanomechanical systems are relatively simple. Highly regular structures provide an obvious set of examples. Among more complex structures, mechanical nanocomputers are of essentially the same complexity as microelectronic computers, once a modest library of logic-element designs has been accumulated (indeed, some CAD software for digital logic design would be applicable without modification). Regular arrays of nanocomputers with local interconnection will not be substantially more complex than single computers, hence the design of extremely complex systems can eventually rely on such arrays to solve computationally expensive problems.

Small molecular manufacturing systems are comparable in complexity to modern automated factories, both in parts count and in organization. Since macroscopic automated factories fall within the range of present human design capabilities, their nanomechanical equivalents should as well. The focus in this chapter on large systems of great complexity should not be taken to imply that *all* useful manufacturing systems need be so complex.

Even the most complex systems described here are arguably less complex than some modern software systems. Although their part counts greatly exceed the number of lines of code in any program, parts and lines of code are not analogous. Mechanisms in a manufacturing system will appear in many copies; a particular mechanism (and its parts) is less like a unique subroutine or object class (and its code) than like one of many invocations of that subroutine, or instances of that object. Indeed, it is often like an invocation or instance *with variables identical to those already used,* existing in multiple copies in order to produce identical "constants," or building blocks, for repeated use in a physical product. Writing a program with 10^{15} lines of code would be out of the question today on grounds of complexity. Running a program that executes 10^{15} instructions is less extraordinary, amounting to a month of supercomputer time. The difference between these cases lies in the massive repetition of standard operations. Nanomechanical systems can exhibit a similarly massive repetition of physical structures, without necessarily entailing unprecedented complexity.

CONCLUSIONS

Molecular mill and manipulator mechanisms can be combined in architectures capable of transforming solutions of small organic molecules into complex macroscopic artifacts built of diamondoid materials and containing components of the sorts. A 1 kg system can build a 1 kg product object in ~1 hr, dissipating ~1.1 kW of heat. The range of products that can be produced is large, encompassing high-performance structures, massively parallel supercomputers, and additional molecular manufacturing systems. A review of costs (other than those of design, licensing, taxation, and the like) indicates that the cost of producing objects using such systems can approach the cost of the required feedstocks, for exoergic reactions. Assuming present materials prices, the anticipated product costs would be ~0.10 to 0.50 $/kg. The complexity of macroscopic molecular manufacturing systems will require extensive automation of the design process.

Some open problems

Far more detailed system-level descriptions of molecular manufacturing could be provided by studies that first attempt to clarify the relationship between (1) the number of distinct building-block types and (2) the range of capabilities that can be delivered by products made from those building blocks. Alternative architectures are worth considering, including architectures that incorporate greater flexibility and more extensive use of general-purpose mechanisms. It can be regarded as a partial survey of open problems in automating the design of products and the specification of production processes. The literature on analogous macroscale problems is a natural point of departure for work in this area.

7

Paths of Molecular Manufacturing

In a *forward chaining* search (as the term is used in computer science), one pursues a goal by taking steps that may lead toward it, sometimes exploring blind alleys. If, however, all possible destinations are considered equally good, then there is no real goal, hence forward chaining never fails. In science, it is common to pursue experimental programs opportunistically, choosing next steps based on immediate prospects and a sense of what is interesting, important, and fundable. This process resembles forward chaining with abundant, unranked goals, and it routinely produces incremental advances in knowledge and capabilities. In *backward chaining*, one first describes a goal, then searches for intermediate situations one step removed from the goal, then for situations one step removed from those, and so forth, planning backward toward situations that are immediately accessible—that is, toward potential first steps on an implementation pathway. If there are many potential first steps, then backward chaining can be particularly attractive. In technology, it is common to select a goal based on its near-term feasibility and economic attractiveness and then plan backward to select the necessary parts and procedures. The enormous range of modern technological capabilities often provides many possibilities, hence engineers are more often concerned with cost and performance than with feasibility alone.

Evaluating Paths to Molecular Manufacturing

Molecular manufacturing is a technological goal (presenting many scientific challenges), but it cannot be achieved in a few steps or a few years. Accordingly, one cannot expect to succeed by combining just existing parts and procedures. Backward chaining is still appropriate, but the links in the chain are intermediate technologies, not mere parts and assembly procedures.

Opportunistic, forward-chaining research strategies have moved toward increasing molecular control, and continued progress will eventually encompass the methods of molecular manufacturing (unless other methods can deliver equivalent capabilities). These opportunistic steps have been guided by perceived short-term rewards, but molecular manufacturing can be achieved more directly if long-term strategies and rewards guide short-term decisions. In a society that heavily discounts its future, however, long-term research directions must deliver short-term rewards.

Table 7.1. Differences between proposed generations of mechanosynthetic assembly systems.

Characteristic	*Stage 1*	*Stage 2*	*Stage 3*	*Stage 4*
System size	macroscopic	submicron	submicron	up to macro
Initial production	present tech.	Stage 1 syst.	Stage 2 syst.	Stage 3 syst.
Molecular parts	$\sim 10^1$ to 10^2	$\sim 10^5$ to 10^6	$\sim 10^7$ to 10^8	$\sim 10^8$ to 10^9+
System assembly	surface chemistry	folding block-chain	positional assembly	positional assembly
Structural materials	folded polymers	crosslinked polymers	crosslinked polymers	diamondoid solids
Instructions	external	external	external	internal
Internal compo	no	no	no	yes
Cycle times	~1s	~100 ms	~10 ms	~1 μs
Error rates	$\sim 10^{-6}$	$\sim 10^{-9}$	$\sim 10^{-12}$	$\sim 10^{-15}$
Working medium	solution	pure liquid	pure liquid or vacuum	vacuum
Reagent transport	diffusion	minimal conveyance	simple conveyance	conveyance
Stray molecules	solvent,	solvent, reagents	solvent, contaminants	rare contaminants
Feedstocks	complex	complex	complex	simple
Reagent processing	none	reactive activation?	reactive activation?	extensive
Reagents	moderate reactivity	high reactivity	high reactivity	extreme reactivity
Applied forces	small	moderate	moderate	large
Energy efficiency	negligible	low	low	good

Short-term rewards aside, the chief criteria for development paths are directness and reliability. A path with fewer sequential steps can likely be traversed more rapidly. A path (or family of paths) that offers many choices at each step is more reliable, since it is unlikely that any single problem can block all choices. Assuming adequate short-term rewards, the total cost of following a development path is of little significance.

SMALLER, SIMPLER SYSTEMS

Macroscopic via Microscopic Manufacturing Systems

With suitable instructions, the exemplar manufacturing system could build a similar system of its own size, or larger. Conversely, it could be built by a similar system smaller than itself. What is the smallest, simplest system of this kind, able to build similar systems that are larger and more complex? Since both CPUs and manipulators can be of submicron size, an initial system can evidently be of microscopic dimensions, but more radical simplifications are possible.

One strategy for simplifying early nanoscale systems is to transfer complexity to the macroscopic world. It is also described how onboard computing, information storage, and power supply systems can be replaced by transducers that respond to externally imposed acoustic signals. It describes how feedstocks consisting of pure solutions of large molecules can enable one-step mechanisms to replace complex molecule acquisition and purification systems. The tactics for building simpler manipulators and for achieving the most important benefits of vacuum operation in a fluid medium have also been described.

Acoustic Power and Control

Broadcast instructions

Replacing computers and stored instructions with broadcast instructions can simplify molecular manufacturing systems. If all operations are sufficiently reliable, then sensors and logic for directing conditional operations can be omitted. (Reliability will result from some combination of computational modeling, model-error-tolerant design, experiment, and redesign.) Broadcast instructions can enable system replication without internally stored instructions.

Signals can be broadcast to molecular mechanical systems in several ways (e.g., by light via photochemical isomerization), but signaling by modulated pressure via mechanical transducers is both simple and adequate. It is thus of interest to consider pressures and pressure histories that can be produced by the application of present technologies.

External control of the pressure history

The pressure on a device can vary over (at least) the range in which the surrounding solution remains liquid. At ~300 K, the solidification pressures of several pure solvents are >2 GPa (for example, >3 GPa for methanol); using mixed solvents can expand the available range. The following calculations assume that pressure can vary over the range $0 \leq p \leq 2$ GPa.

Available laboratory equipment can subject 50 cm³ of solution to static pressures of up to 2 GPa. Broadcasting of instructions, however, requires varying the pressure. At low frequencies, suitable designs include small cells modulated by pistons driven by relatively large actuators. If the diameter of the cell and piston are ~10^{-3} m (permitting volumes $>10^9 \mu^3$), then the maximum required force is ~2×10^3 N, and electromechanical actuator dimensions <5 cm appear reasonable. It appears feasible to sweep the ambient pressure in a cell through its full range in ~1 ms. The approach outlined below uses the ambient pressure in the cell to select transducers of a particular class, and acoustic pressure fluctuations of lesser magnitude (± 1 to 100 MPa) and greater frequency (~10^6 to 10^7 Hz) to drive them. (Acoustic wave amplitudes can be increased by focusing.)

At pressures up to 2 GPa, the speed of sound in organic liquids is ~1000 to 2000 m/s, with negligible dispersion of phase velocities at megahertz frequencies. The amplitude of acoustic plane waves in a liquid decays exponentially with distance, and is usually well described by the expression

$$A = A_0 \exp\left(-C_{atten} f^2 x\right) \quad ...(1)$$

where A is the amplitude, A_0 the initial amplitude, f the frequency (Hz), x the propagation distance (m), and C_{atten} a measure of the attenuation (s²/m). For organic liquids at ~300 K and low pressures, the value of C_{atten} ranges from ~3×10^{-14} to ~6×10^{-12} s²/m, declining with increasing pressure. For a typical value, $C_{atten} = 10^{-13}$ s²/m, a 10 MHz signal can propagate ~1 cm with ~10% attenuation; sizes and frequencies in this range are compatible.

A final concern is conversion of high-amplitude sound waves to shock waves by nonlinear properties of the fluid. Shock-wave formation, however, requires a substantial propagation distance and does not occur in systems having sizes, frequencies, and amplitudes in the range considered here. At 10 MHz, a 100-pulse train lasts 10^{-5} s, during which the ambient pressure (if modulated over its full range in 1 ms) changes by only ~1 % of its peak value. It will here be assumed that, at a particular point in the cell, the pressure can be varied over a 2 GPa range in 1 ms, and that acoustic pulse trains with frequencies of up to 10 MHz can be superimposed. For comparison, pulse-echo diagnostic ultrasound systems used in medicine can produce peak-to-trough pressure differentials of ~10 MPa, with frequencies of 1 to 20 MHz and pulse lengths measured in microseconds.

Pressure-threshold actuators

To convert a pattern of pressure fluctuations directly into a series of mechanical operations requires transducers that function as pressure-driven actuators. Devices prove convenient. Each actuator of this kind is characterized by a pressure threshold p_{thresh}, a volumetric change ΔV, and a corresponding length change Δl. The Gibbs free energy difference between the extended and retracted configurations is

$$\Delta G_{extend} = (p - p_{thresh})\Delta V \quad ...(2)$$

and the outward force applied by the actuator at lengths between its limiting values is

$$F_{extend} = -(p - p_{thresh})\Delta V / \Delta l \quad ...(3)$$

neglecting the effects of compliance. By adding a mechanism such as a lever, the effective value of Δl can be changed without varying other parameters. The discussions that follow will use one of two sample sets of parameters:

1. High-pressure actuator parameters. With $\Delta p = p - p_{thresh} = -0.1$ GPa and $\Delta V = 5$ nm^3, $\Delta G_{extend} = -500$ maJ, providing reliable avoidance of the disfavoured position despite thermal excitation. With $\Delta l = 1$ nm, the maximum force delivered at $\Delta p = -0.1$ GPa is 0.5 nN.
2. Low-pressure actuator parameters. With $\Delta p = -0.001$ GPa, $\Delta V = 500$ nm^3, ΔG_{extend} is again -500 maJ/nm^3, and with a suitable geometry and mechanical advantage, the delivered force can again be 0.5 nN.

In a system described by the high-pressure parameters, a 2 GPa pressure range permits use of actuators divided into 20 distinct classes, with $p_{thresh} = 0.1, 0.2, 0.3, ..., 1.9, 2.0$ GPa. With the low-pressure parameters, 20 classes can be accommodated in a pressure range of 20 MPa, or 2000 in a pressure range of 2 GPa.

Actuators in anyone pressure class can be cycled through a reliable length transition repeatedly without activating those of other pressure classes. Shifting the ambient pressure monotonically from the value of

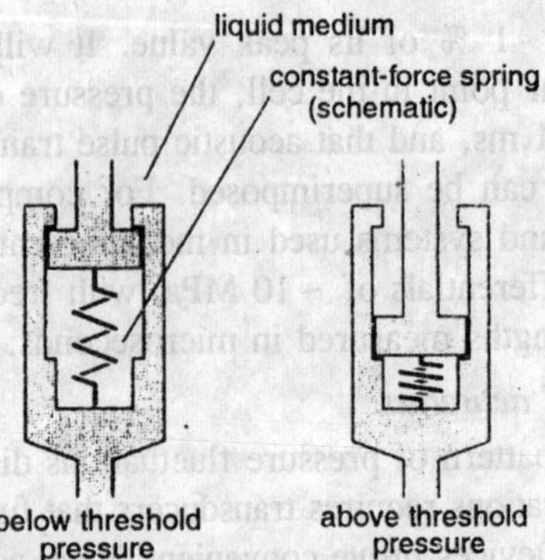

Fig. 7.1. Schematic diagram of a pressure-threshold actuator, here illustrated with a constant-force spring and a piston moving between two limit stops.

p_{thresh} for one class to that of another class, then back again, will cycle all actuators of intermediate p_{thresh} exactly once. If the pressure is swept over the range only once, each set of actuators becomes active only once, each in a fixed sequence but with an independently controllable number of cycles including at least one more extension than retraction.

Control via acoustic signals

Two actuators with different thresholds could provide clock and data signals, permitting the input of strings of bits to a nanomechanical controller with data rates on the order of 10^7 bits per second. The present application, however, requires control of the motion of several different parts, which would require mechanisms to decode the instructions and then apply the results at different locations.

A more direct approach can reduce complexity and illustrates how pressure-threshold actuators can drive a ratchet mechanism that moves a rod. When the *latching pawl* is engaged, the rod will not move left or right spontaneously, but can readily be driven to the right by an applied force. The *driving pawl* provides this force: every downward pressure transition from p_{thresh} to $p_{thresh} - \Delta p$ pushes the rod one increment to the right. (A reasonable spacing for the ratchet teeth is ~ 1 nm.) When both ratchets are disengaged, a small bias force retracts the rod leftward to a fixed position. For a mechanism of this kind to work reliably, the ratchet barrier heights must be adequate (e.g., 250 maJ); actuators with ΔG_{extend} on the order of 500 maJ are more than adequate to surmount these barriers. Incrementing a ratchet dissipates substantial energy, but efficiency here is not a concern. A typical operational cycle would begin at high pressure, with both

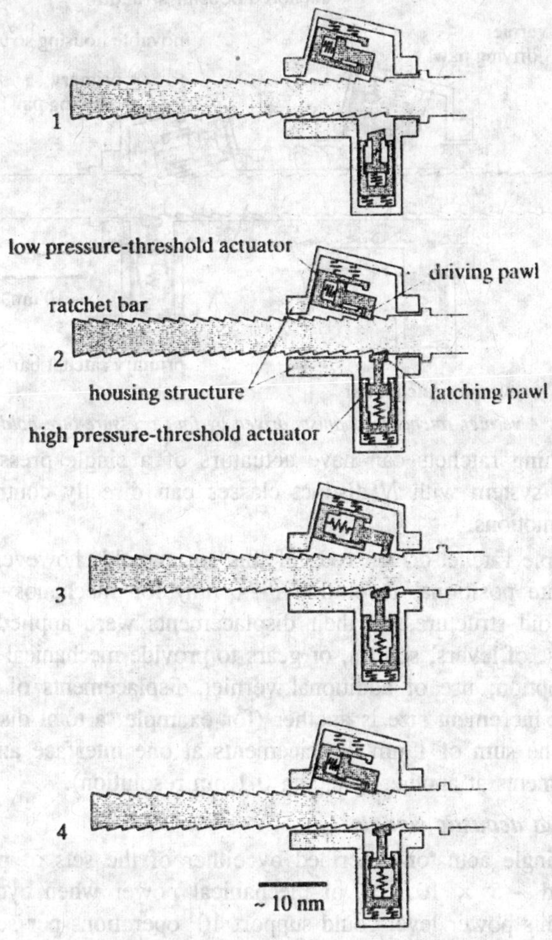

Fig. 7.2. A ratchet mechanism driven by a pair of pressure-threshold actuators.

ratchets disengaged and the rod to the left. Lowering the pressure through the first range engages the latching ratchet with a force that is essentially constant at all lower pressures, owing to the limit stop. After lowering the pressure further (possibly much further), the driving ratchet extends and moves the rod to the right. The pressure then can be lowered further with no effect on this mechanism, or it can be cycled up and down across the active range of the driving ratchet, moving the rod by further increments. After an operation has been performed, a return to high pressure resets the rod to its initial position.

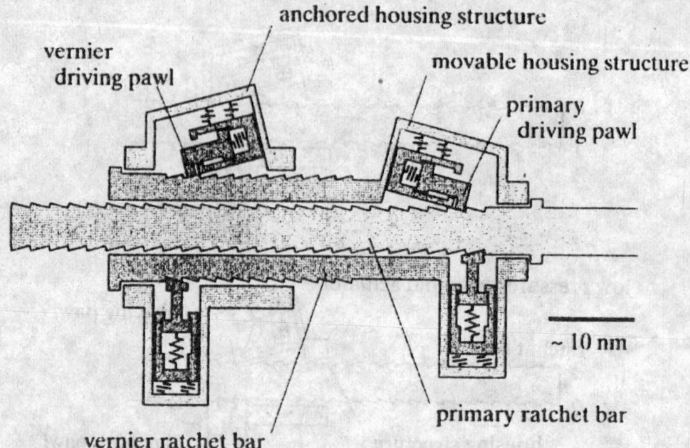

Fig. 7.3. A vernier ratchet mechanism driven by four pressure-threshold actuators.

All latching ratchets can have actuators of a single pressure class, hence a system with N distinct classes can directly control $N - 1$ distinct motions.

Simple ratchet-driven rods of this sort would, however, provide inadequate positional resolution (~ 1 nm) for mechanosynthesis of diamondoid structures, if their displacements were applied in a 1:1 ratio. Use of levers, screws, or gears to provide mechanical advantage is one option; use of additional vernier displacements of a slightly different increment size is another (for example, a total displacement can be the sum of 1 nm displacements at one interface and 1.1 nm displacements at another, yielding 0.1 nm resolution).

Power via acoustic signals

A single actuator described by either of the sets of parameters can yield $\sim 5 \times 10^{-12}$ W of mechanical power when cycled at 10 MHz. This power level could support 10^6 operations per second at 5 aJ/operation. In systems of the sort considered here, control and power supply are inseparable. The power supplied by an actuator is sufficient so long as the forces it applies are sufficient to drive component motions. A force of ~ 0.5 nN is ample to overcome both the inertial and viscous drag forces caused by displacing a 10 nm diameter, 100 nm long rod by 1 nm in 10^{-7} s; these values are chosen to roughly model the struts of a molecular manipulator.

Simpler Manipulators

The manipulator requires stiff materials and intricate parts, making it difficult to implement. An alternative design uses a *Stewart platform*,

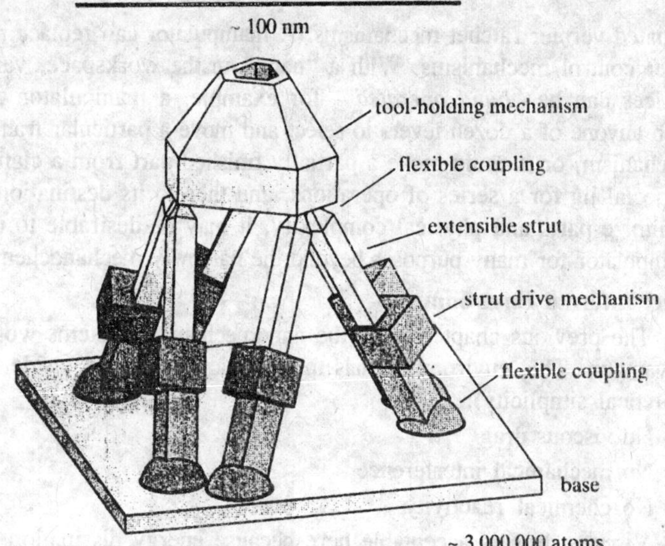

Fig. 7.4. A positioning mechanism based on the Stewart platform.

widely used in aircraft flight simulators and occasionally in robotics (its use in this context was suggested by Professor D. Taylor of Cornell). The Stewart platform has a roughly octahedral geometry, with a fixed triangular base linked to a mobile triangular platform by six struts of adjustable length. Fig. 7.5 shows a two-dimensional analogue with three struts and indicates a portion of the available range of motion. Strut lengths can be altered with ~0.1 nm accuracy by means of pressure-

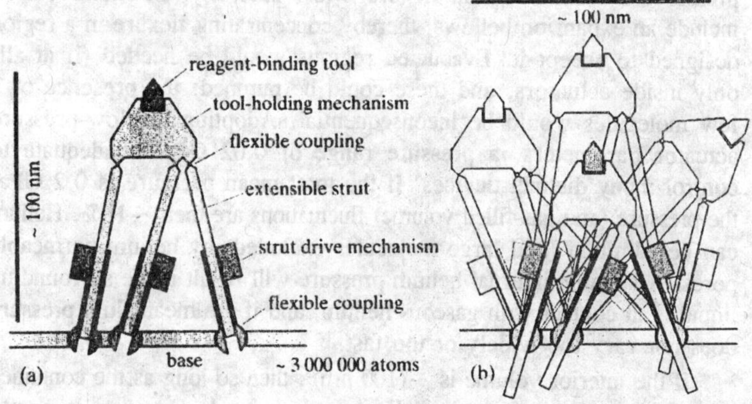

Fig. 7.5. A two-dimensional schematic of a positioning mechanism based on the Stewart platform.

actuated vernier ratchet mechanisms. A manipulator can replace many other control mechanisms. With a "hand" in the workspace, various devices can be "*hand operated*"—for example, a manipulator could push anyone of a dozen levers to select and move a particular transport mechanism, or it could move a partially finished part from a clamp to a special jig for a series of operations, and then to its destination. To minimize parts and physical complexity, it may be desirable to use a manipulator for many purposes beyond the narrowly mechanochemical.

Inert Internal Environment

The previous chapters describe nanomechanical systems working in vacuum. This environment has three basic advantages (aside from analytical simplicity):

1. No viscous drag
2. No mechanical interference
3. No chemical reactivity

Viscous drag is acceptable here because energy dissipation is of negligible concern. Mechanical interference (i.e., jamming) could pose problems, but only if a motion (e.g., of gears meshing) creates an enclosed pocket of molecular size or larger and then attempts to compress it to a substantially smaller volume; mechanisms can be designed so as to avoid this. Chemical reactivity is a concern in many media (particularly with highly active reagents), but not in a nearly inert medium such as helium.

A helium-filled structure can equalize interior and exterior pressures to minimize wall flexure. More precisely, an enclosure can include an expansion bellows, thereby concentrating flexure in a region designed to accept it. Evacuated regions would be needed (if at all) only inside actuators, and these could be pumped: the presence of a few molecules would be inconsequential. Adopting the low-pressure actuator parameters, a pressure range of 0.02 GPa is adequate to control many distinct devices. If the total mean pressure is 0.2 GPa, the pressure (and gas-filled volume) fluctuations are then ~10%. Helium can be admitted (and large molecules excluded) by helium-permeable pores. A suitable internal helium pressure will result if the surrounding liquid is in contact with gaseous helium, and if the mean fluid pressure does not vary too widely or too fast.

If the interior volume is $\sim(100\text{ nm})^3$, then so long as the contained helium in a set of devices has a mean mole-fraction impurity concentration $<10^{-9}$, the probability of the presence of a stray nonhelium molecule in anyone device is $<.1$, hence most lack any contamination. Many reactive contaminants could be neutralized by *gettering*, that is, by providing sacrificial reactive surfaces with which they can combine without harmful effect. One such contaminant is hydrogen, the molecule most likely to enter through helium-permeable pores.

Sorting and Ordering Molecules

To simplify cost estimation and to reduce estimated production costs, molecular manufacturing systems able to use low-purity solutions of small-molecule feedstocks. The present objectives differ; to simplify the systems themselves (and to reduce estimated development costs), it is better to assume larger-molecule feedstocks of high purity.

Large feedstock molecules of several kinds can differ greatly both from solvent molecules and from one another. Reliable acquisition of a large molecule of a particular kind can then be straightforward, in part because larger molecules can have enormous binding affinities. For example, a 31-atom molecule (biotin) binds to a protein (avidin) in water with a dissociation constant K_d of $\sim 10^{-15}$ nm^{-3}. Assuming solute concentrations of ~ 1 nm^{-3}, a dissociation constant this low provides more than adequate sorting and ordering in a single step, assuming that no closely similar molecules are present to compete for the receptor. This can be ensured by careful purification of the large-molecule components of the feedstock solution (e.g., by repeated affinity chromatography). Accordingly, it seems that a judicious choice of high-purity feedstock components can enable a manufacturing device to acquire feedstock molecules of one kind with good reliability in a single step. A mechanism with a single receptor could then replace complex systems.

Minimal Diamondoid-material Systems

Any attempt to describe a "*minimal*" system must be considered speculative. On one hand, a general description may overlook some minor but complex and necessary feature; on the other hand, clever designs may result in smaller, simpler systems overall. The mechanosynthetic system described here has passed through several cycles of simplification, and there is no reason to regard it as ideal or final. This sketch, like those that follow for other stages, is offered only to provide a stimulus and point of reference for further work.

Preparing and applying reagents

This section describes systems using specialized mill mechanisms to prepare reagents and to perform several stages of subsequent assembly; this specialization improves throughput and energy efficiency.

Parts count can be reduced, however, by applying manipulator mechanisms more directly to feedstock molecules.

Conventional chemical synthesis can prepare feedstock molecules of several kinds, each having a reagent-precursor moiety at one end and a distinctive "handle" of several dozen atoms at the other. Differences between handles can enable the selective binding. Forces applied by a manipulator can drive the rotation of a receptor from the exterior to the interior of the enclosure, and can force the release of a bound feedstock molecule; the handle can then bind to a larger block with a standardized, grippable surface.

A pair of manipulators (one can be an immobile gripper) can be situated so that their tips can interact. Bound feedstock molecules can then be moved, placed, and made to interact with flexible control of encounter sequence and geometry. The resulting mechanochemical transformations can generate a wide range of reagent moieties and subassemblies from a smaller range of feedstock molecules. In particular, relatively unreactive precursors (stable in solution) can be transformed into highly active reagents in the inert environment within the enclosure.

Waste disposal

The preparation and application of reagent moieties will generate wastes in the form of discarded handle structures. These will typically have a volume greater than that of the product structure, and must be cleared from the workspace. The manipulator can deliver them to a relatively unselective positive-displacement pumping mechanism for expulsion into the surrounding solution.

Spent handles might resemble feedstock molecules enough to compete for receptors. This problem can be avoided by either (1) ensuring that their use modifies structures important to binding, (2) ensuring that their concentration in solution remains very small, or (3) trapping them in an expandable external compartment (e.g., a bag).

Wall design and product delivery

Pressure equalization minimizes loads on walls, enabling the use of a flat-walled enclosure scheme. This geometry relies on a sliding interface to provide a seal between the enclosing wall of the manufacturing system and the emerging wall of the (possibly larger) product structure.

Estimated size, complexity, and replication time

A system of the sort described in this section could be packaged in a volume ~(100 nm)^3 (although not equidimensional). The structural components of a Stewart-platform manipulator having 6 struts ~100 nm long and ~5 nm wide contribute ~2×10^4 nm^3 of solid structure. Each strut has 4 actuators, including verniers. If an additional 4 actuators are provided to control operations at the manipulator tip, 8 to control a second, more rudimentary arm, 4 to control overall movement of the product structure, and 1 to drive a waste-expuhion pump, then the total is 42. If each actuator is approximated as a box 10 nm on a side with walls 1 nm thick, actuators contribute ~2.5×10^4 nm^3 of solid structure (actuators this large permit operation in the low-pressure regime. Allowing for 10 reagent-acquisition mechanisms, each ~5 nm wide and ~10 nm tall, adds a negligible 0.25×10^4 nm^3. Walls 1 nm thick contribute ~6×10^4 nm^3. The total volume, including a 50% allowance for additional components and structures, is then ~1.6×10^5 nm^3, or ~2×10^7 atoms. The estimated number of moving parts is, according to this estimate, ~100. With the parameters assumed here, motion frequencies are limited by the frequency at which the pressure in the cell is swept over its full range. At ~1 ms per sweep and 5 motion cycles per reagent application cycle, reagent application operations can be performed at ~200 s^{-1}. If the mean number of atoms transferred per operation is ~1, then the replication time for the system is ~10^5 s, or ~1 day. Since there is reason to regard the atom count as generous and the motion frequencies as conservative, this time estimate seems likely to be generous.

Instrumentation

Techniques for probing product structures will be central to experimentation and process development. One approach would exploit modulation of the fluorescence yield, lifetime, and spectrum of a chromophore by changes in its mechanical and electronic environment. In the solution phase, changes in solvent polarity can produce dramatic shifts in fluorescence spectra. Changes in the distribution of charge near a bound chromophore or direct mechanical deformation of a conjugated structure can produce changes of similar magnitude.

Existing instrumentation using subnanosecond excitation pulses and spectrally filtered, nanosecond time-gated detection to suppress background photons permits the detection of single molecules (each with a single chromophore) by their fluorescence in a ~10 ms interval. With accessible improvements (e.g., reduction of photobleaching by tailoring the molecular environment) more photons can be collected, and changes in the properties of the chromophore should be readily detectable. This can enable readout of data from an individual mechanosynthetic device. A chromophore can be manipulated like a feedstock molecule, and its fluorescent properties (frequency, quantum yield) can be modulated by direct contact with the structure being probed, causing electrostatic or mechanical perturbation of the chromophore. Since the chromophore is a proximal probe, repeated positioning at different locations can be used to build up an image of a workpiece structure, as in AFM and STM. If measurements can be performed in ~10 ms or less, atomic-resolution images can be acquired at ~1 nm^2/s.

Unless their photostability is excellent, probe chromophores will require regular replacement. This can be accomplished by mechanisms like those used to provide reagent moieties. To avoid interference from unbound chromophores, however, the feedstock molecules must initially have negligible fluorescence at the signal wavelengths, and must be activated after binding to the mechanism. Alternatively, the mechanosynthetic device could be tethered to a surface and fluid flow could be used to purge the surrounding volume before gathering data from a proximal fluorescent probe.

Summary

An inert workspace environment in which highly active reagent moieties can be positioned relative to workpieces with high precision and reliability (the essential requirements of qualitatively advanced molecular manufacturing) can be provided by a mechanism with ~10^2 moving parts and ~2×10^7 atoms. Pressure-threshold actuators enable substantially independent control of several tens of mechanisms within a single manufacturing system; an indefinitely large number of identical systems can be driven through the same sequence of operations at the same time. Using a manipulator mechanism able to reach a large portion of the interior space, complex behaviours can be implemented by complex control sequences, rather than by complex hardware. With 10^7 Hz actuator stepping rates, manufacturing can proceed at ~200 operations per second, enabling a system to build an object of its own complexity in about a day. Molecular probes can be used to determine the outcome of operations within a single device, facilitating experimentation and the development of new operations.

These system characteristics assume the use of diamondoid materials, and are chosen to permit the fabrication of diamondoid materials. The following steps in the backward-chaining analysis will move away from these materials and conditions toward more familiar polymeric structures and solution-based chemistry.

Softer, Smaller, Solution-phase Systems

Diamondoid via Nondiamondoid Systems

Diamondoid materials with diamond like stiffness and atom number density are challenging targets for synthesis, and are here assumed to require the use of highly active reagents. These, in turn, are assumed to be incompatible with solution-phase synthesis conditions, owing to the potential for reaction with solvents. (If either of these assumptions proves incorrect, development will be easier.) Along development paths that start with solution-phase chemistry, diamondoid structures must be built by nondiamondoid systems.

A nondiamondoid system can build diamondoid structures if it can execute suitable motions with adequate stiffness in an environment compatible with highly active reagents. A mechanism resembling will suffice, using thicker struts to compensate for lower modulus. Highly oriented polymers can exhibit a Young's modulus > 100 GPa; examples include polyethylene and Kevlar 49. Similar materials can be made by solution-phase mechanosynthesis. A strut ~ 10 nm in diameter and ~ 100 nm long would then have a stretching stiffness $k_s \approx 100$ N/m.

Highly crosslinked organic structures of good stiffness can presumably be made by positioning multifunctional monomers of conventional reactivity; extensive use of cycloaddition reactions, for example, can enable the formation of relatively dense bond networks. The number of atoms required for a given functional capability will usually be larger for such structures than it would be for more diamondlike structures, but the number of (relatively large) monomers placed, and hence the number of mechanosynthetic operations, should typically be lower.

These crosslinked organic structures can be built by mechanosynthetic systems of still lower stiffness. Required manipulator stiffness varies as the inverse square of the site separation. The large size of multifunctional monomers in comparison to simple radicals and alkynes tends to increase the separation between reactive sites; the alignment requirements of cycloaddition reactions can further widen the separation in configuration space.

The picture that emerges from these considerations is of a progression from relatively compliant mechanosynthetic systems that position large monomers in solution to relatively stiff systems that position small, highly active moieties in an inert environment. Monomer-based structures built by compliant mechanisms using solvent-compatible chemistry can themselves be stiff enough to build diamondoid structures.

Inert Environments from Solvent-based Systems

The transition from solvent to an inert environment can be achieved by a manufacturing system that (1) tolerates but does not require an internal solvent, and (2) can replicate while maintaining a barrier against solvent entry. Consider a system meeting this specification that initially contains 10^9 molecules of solvent. After one replication cycle, the mean number of solvent molecules in each of the resulting pair of systems is $10^9/2$. After 40 replication cycles, the mean number is $10^9/2^{40} \approx 10^{-3}$; hence most systems are solvent-free by virtue of extreme dilution.

If all solvent molecules were sufficiently mobile, then replication-based dilution could be replaced by simple pumping. Problems that might be caused by operating in the presence of a helium-solvent phase boundary can be avoided by pumping out all mobile solvent molecules, then using replication cycles to produce solvent-free mechanisms by dilution.

If enclosures made of polymeric materials are permeable to small reactive contaminant molecules (e.g., hydrogen), then either the external concentration of such contaminants must be kept low (by avoiding sources or providing sinks) or the internal mechanisms must tolerate their presence. Note that dangerously reactive contaminants should be susceptible to chemical scavenging.

Solution-synthesized Pressure-threshold Actuators

Relatively advanced solution-phase mechanosynthesis systems can presumably make (and use) pressure-threshold actuators based on piston mechanisms. An alternative approach, presenting less synthetic difficulty, exploits pressure-driven changes in volume analogous to solidstate phase transitions in bulk materials. Various salts and organic crystals exhibit phase changes at pressures between 0.05 and 2 GPa with fractional volumetric changes of ~ 0.05 to 0.1. A block of material with this behaviour can serve as a pressure-threshold actuator: the nature of the relationship among force, displacement, and pressure is identical. The chief disadvantage is the smaller value of $\Delta V/V$; this can be accommodated by roughly doubling the linear dimensions of the device.

Smaller Liquid-based Mechanisms

It is assumed that synthesis occurs inside a ~ 100 nm scale enclosure, with each moiety (or monomer) emplaced by a manipulator of comparable size. The enclosure must surround the system, and the manipulator (to enable replication) must span its diameter. A rough estimate places most of the system mass in these components. To add a further link in the backward chain from advanced manufacturing systems, it is desirable to keep the chief benefits of enclosures (exclusion of stray reagent molecules) and of manipulators (use of positional control to build complex structures) while eliminating most of their mass. The result is an easier target for construction.

Stray reagent exclusion without a wall

If reagent monomers contain dozens of atoms (or have discardable handles), then it is not a severe constraint to require that they carry enough polar or charged groups to make them substantially soluble only in a polar medium. If the workspace is located within a droplet of an unreactive, nonpolar solvent (e.g., an aliphatic hydrocarbon), then reagent monomers carried by an external polar solvent will be largely excluded from the interior. Droplets can be stabilized against coalescence by polymeric surfactant structures of low mass and complexity; their equilibrium size can be fixed (for example) by the colligative properties of their contents, or by the surfactant structures. Reagent monomers can be transported into the workspace by mechanically rotatable receptors.

Folding-assisted assembly

Biological molecular machines are manufactured by devices capable of controlling only the sequence of a chain of small monomers. A manipulator-based assembly system with even a small range of motion (e.g., 5 nm) could build relatively rigid, highly crosslinked molecular objects of substantial size—more like a protein than like an amino acid residue. A chain of these objects could be designed to undergo a specific, predictable folding sequence as it exited the assembly system, guided by the binding of complementary surfaces with areas of several square nanometers. Note that identical pairs of complementary surfaces can be used repeatedly, so long as only one of any particular kind is ever exposed within reach of a tethered complementary structure.

Accordingly, a manipulator with a range of motion ~1/20 can build structures far larger than itself, including larger manipulators. A single ratchet-driven bar can be used to move any of several reagent receptor mechanisms within reach. As in the proposed AFM-based mechanism, the workpiece can be mounted on the manipulator, enabling monomers to be bound to it directly from a receptor.

The advantages over an AFM-based mechanism include (1) isolation of the workpiece from stray reagent monomers, (2) motion control in six degrees of freedom, (3) a "substrate" that can include mechanisms for product clamping and release, (4) reagent receptors in known locations that need not be mapped, and (5) the feasibility of directing a replication process that generates large numbers of low-cost mechanosynthetic devices. The disadvantages include (1) inability to correct for reaction failures—at least when many devices are sharing one stream of broadcast signals—and (2) the requirement for much more complex nanomechanisms.

Estimated size

To minimize the scale of the pressure-threshold actuators, pressures and sizes can be chosen in the range suggested by the high-pressure actuator parameters. With $\Delta V = 5$ nm^3, the required volume per actuator is ~50 nm^3, and a set of ~40 actuators (of ~20 different pressure classes) requires ~2000 nm^3 of material. The size of the struts is negligible in comparison, and the wall is nonexistent. In systems with large pressure swings, structural deformations resulting from ordinary compressibility (rather than discrete phase changes) will be an important consideration in design.

Based on these estimates, a total structural volume of ~4000 nm^3 appears ample, suggesting an atom count of $\sim 4 \times 10^5$. Assuming ~20 atoms per monomer, the number of synthetic operations required for construction is $\sim 2 \times 10^4$. This is within the $>10^5$ operations estimated to be feasible using AFM-directed mechanosynthesis; alternatively, a self-assembled structure might be made from a set of ~400 molecular objects each consisting of ~50 monomers. Once fully operational, a device of this sort could build another device of comparable size and complexity in ~1 hour, working at only 5 operations per second.

Development Time

The preceding sections outline an ambitious effort. Reaching the first stage described in this chapter will require the construction of precise molecular structures ~10^2 to 10^3 times larger than those previously achieved. Subsequent steps, building on this base, will move still farther from present laboratory experience.

Naive assessments of feasibility often implicitly assume that *future* developments will be limited by *today's* tools, instruments, and techniques. This mistake is natural, because only today's capabilities are widely and concretely known. Where the developments under consideration involve several stages of tool development, however, this static picture can be quite misleading. To formulate a more accurate understanding, one must instead consider each step in light of the capabilities provided by previous steps. The balance of this section makes a preliminary effort in this direction, beginning with a conceptual framework of more general applicability.

Determinants of the Development Time

Cycle of experimentation

Development of new technologies involves experimentation leading to the accumulation of reliable systems and tested capabilities. The cycle of experimentation has several stages: (1) choosing an objective, (2) choosing an approach, (3) trying the approach, and (4) evaluating the result. So long as the result is unsatisfactory, stages (2) through (4) are repeated until success is achieved or the objective is altered or abandoned.

Number of sequential experimental steps

The time required to develop a system equals the sum of the durations of the required sequential steps. Good strategy can reduce the number of steps (and their difficulty); there is no reason to believe that the best strategy for developing molecular manufacturing has yet been identified. Further, many steps can be taken in parallel. For example, multiple teams can work with similar mechanosynthetic systems, each doing experiments that add to the repertoire of reproducible synthetic operations. Low instrument and materials costs can enable more teams to work, increasing parallelism. Concurrent development of a system and of plans for using it (via computational modeling) can further increase parallelism.

Number of teams working on a step

Even without cooperation, work by multiple teams can speed completion of a single step, because this happens as soon as the fastest (or luckiest) group succeeds. The number of teams funded to work on a problem will depend in part on the perceived importance of solving it.

Computational modeling

If computational modeling were perfect (fast, inexpensive, and accurate), then it would enable all correctly executed experiments to succeed on the first try. In reality, it does far less, yet modeling can help researchers avoid unworkable approaches without rejecting too many workable approaches. This tends to increase the fraction of experiments that can, after repeated tries, be made to work, thereby decreasing the mean delay in achieving an objective.

Computational modeling is more useful in planning mechanosynthesis than it has been in solution synthesis. Mechanosynthesis can impose a particular interaction geometry on a pair of reagents, thereby defining the system to be modeled and rendering small errors in transition-state energy relatively unimportant. Further, relatively rigid, crosslinked structures are less sensitive to modeling errors than are conformationally flexible peptides and typical products of organic synthesis. As inexpensive computers become faster and software becomes more sophisticated, the utility of computational modeling will increase.

Time for synthesis

If the cycle of experimentation is to be short, the time required to synthesize a structure must be short. This consideration favours active reagents, high effective concentrations, and the use of software control to direct long sequences of operations.

Time for evaluation

Short experimental cycles likewise demand short evaluation times. This tends to favour intermediate technologies that combine sensors and manipulators in a single package.

Willingness to limit evaluation and abandon subgoals

In routine science, an unexpected result often provides a problem to study. In building novel systems, however, an unexpected result often gives reason to seek a more predictable system. To evaluate a result, success and failure must be distinguished, but the nature of a failure need not always be understood—an attempt to study it in detail seldom yields fundamental knowledge, and may merely yield detailed knowledge of something useless. Even a consistent pattern of failures may motivate, not a prolonged halt for study, but instead a redesign that avoids the difficulty entirely. Analyzing unexpected failures can be of great value, producing rules to guide future design; but quickly trying a different approach can sometimes speed development. (Note that evolution developed sophisticated molecular machine systems without analyzing failures.)

Brownian Assembly of Medium-scale Blocks

Building a solution-phase mechanosynthesis system of the sort by Brownian assembly (rather than by AFM-directed mechanosynthesis) might involve developing ~400 self-assembling building blocks of ~50 monomers apiece. Monomers can be developed in parallel with one another. With good CAD software to define their functions and interfaces, building blocks can also be developed in parallel with one another. Further, design of building blocks can be overlapped with the design and synthesis of monomers.

Designing ~400 building blocks need not require the design of a similar number of unique complementary interfaces. If building blocks are added to solution sequentially (and removed afterward), each will have a unique destination so long as the assembly sequence exposes only one interface of each kind at any one time; supramolecular variants of protective-group strategies may be useful here. The quality of the tools available for computer-aided design of folded structures that meet geometrical and functional specifications appears critical to the pace of development along this pathway.

Blocks of ≥150 monomers can be characterized by NMR spectroscopy. Interpretation of NMR data is simpler if the desired fold is known, and if structures that differ widely from the objective are discarded without detailed study. Molecular-tip AFM techniques could aid in evaluating both building blocks and assembled structures.

In past protein design efforts, the time required for synthesis and characterization has often been measured in months or more, and success rates per trial have been moderate. Augmented proteins, designed to make design and synthesis easier, will presumably permit somewhat faster cycles.

Mechanosynthetic Assembly of Small Building Blocks

AFM-directed mechanosynthesis can sidestep the problem of designing folding polymer structures and enable immediate characterization of product structures using the AFM mechanism itself. The required monomers are likely to be complex (bearing multiple functional groups) but can be designed to permit synthesis by conventional means. Monomers can be developed in parallel, and to avoid investing synthetic effort in chemical blind alleys, they are best developed in parallel with designs intended to use them.

Effective exploitation of AFM-directed mechanosynthesis systems will require substantial software development: image interpretation software able to determine the types, positions, and orientations of specific sets of reagent binding molecules; control software to automate reagent positioning and sensing, thereby automating the execution of long sequences of reactions; and computeraided design software to plan both what to build and how to build it.

The cost of an AFM -based mechanosynthesis system can be estimated today: it will likely be less than the price of an AFM system in the early 1990s, ~$\$10^5$. Reasonably low prices will enable many teams to enter the field, some working in parallel to develop and expand a set of reliable synthetic operations. With suitable reagents, reaction times can be ~1 s, enabling the construction of moderately complex objects in minutes and of objects containing $\sim 10^5$ monomers in about a day. Once good tools are in hand, the time required to develop mechanical systems with $\sim 10^2$ moving parts consisting of $\sim 2 \times 10^4$ monomers (estimated to be sufficient to build a minimal solution-based molecular manufacturing system) could be a fraction of a decade.

First-generation Solution-based Systems

Manufacturing systems have capabilities resembling those of AFM-based me chana synthetic systems, hence they can use identical or similar monomers, and their applications can be planned using similar CAD software. Since these systems are precisely constructed, they create no requirement for image-interpretation software to locate and discriminate among reagent receptors. Evaluation of the results of an operation (including imaging) can be achieved.

The estimated time required for a system of this kind to build a copy of itself (a key threshold in development) is several hours. If the replication time is ~1 hour, then the time required to produce a macroscopic quantity of systems is several days. Subsequent production of other devices can supply relatively low-cost, disposable laboratories for mechanosynthesis. The cost of required reagents is negligible on a per-device basis. For example, a cost of 10^4 \$/gm amounts to $\sim \$10^{-12}$ per complex device. New reagents, operations, and devices can be developed in parallel in many laboratories. From this technology base, the direction of advance is toward enclosed systems with larger manipulators, then toward inert interiors, then toward more active reagents. Experimental cycle times can remain short throughout this process.

Inert Environments, Diamondoid Materials

After the transitions enabling the use of highly active reagent moieties and the construction of devices made from diamondoid materials, development can proceed more or less directly to advanced molecular manufacturing systems. CAD systems for diamondoid materials will differ from those used with larger monomers, but will be in some respects simpler. The frequency of synthetic operations can be increased, in part by the use of more active reagents and in part by the construction and use of decoding subsystems that more fully exploit the capacity of information input channels. Further, once systems are built with internal control and data storage devices, brief instructions can activate complex subroutines, greatly relaxing constraints associated with data transmission rates. A single manufacturing system can then contain multiple production lines, each working at high frequency. Construction of molecule sorting and preparation subsystems can enable the use of simpler, less pure, less expensive reagents, reducing product costs. With scale-up, integration, and suitable software for design and control, these features provide the basic capabilities.

Conclusions

A backward-chaining analysis indicates that feasible developmental pathways link our present technology base to the technology base described already. Alternative pathways are known, and others can be devised. To reach the first stage discussed in this chapter requires the assembly of molecular machine systems from ~ 10^4 to 10^5 monomers. It describes several ways to achieve this ability, via Brownian assembly or the development of AFM-directed mechanosynthesis systems. Once assembled, small mechanosynthetic devices can be made to execute complex patterns of motion powered and controlled by imposed pressure fluctuations. Relatively small and simple devices can be used to construct larger and more complex devices. A series of steps can enable a relatively smooth transition from solution-phase assembly of monomers with conventional reactivity to the assembly of diamondoid mechanisms in an inert (eventually, vacuum) environment using highly active reagents.

Each step along this pathway will present great practical challenges, but each step will also bring valuable new capabilities. The long-term rewards, measured in terms of scientific and technological capabilities, appear large.

8

MOLECULAR DYNAMICS

NONSTATISTICAL MECHANICS

Molecular dynamics is fundamental to molecular machinery and has received extensive study in physical chemistry and chemical physics. This section outlines several nonstatistical descriptions of molecular dynamics used in scientific work, commenting on their applicability to nanomechanical engineering problems.

Vibrational Motions

Molecular vibrations have been extensively studied in connection with infrared spectroscopy, and observed infrared vibrational frequencies are a major constraint used in determining parameters for molecular mechanics energy functions. Small displacements of a stiff structure from its equilibrium geometry are associated with nearly linear restoring forces. This permits the use of a harmonic approximation, in which the vibrational dynamics can be separated into a set of independent normal modes and the total motion of the system can be treated as a linear superposition of normal-mode displacements. This common approximation is of considerable use in describing nanomechanical systems.

Since both classical and quantum mechanics permit exact solutions for the harmonic oscillator, the time evolution of systems can readily be calculated in the harmonic approximation. In real systems, nonlinear effects permit energy exchange among vibrational modes, causing relaxation to thermal equilibrium. The equilibrium state, in turn, is best described by the methods of statistical mechanics. A nonstatistical description of vibrational motion is of interest chiefly during (or within a few relaxation times of) the excitation of a vibrational mode by a nonthermal energy source.

Rate of Reactions

Molecular beam experiments

Vibrations involve motion within a potential well; reactions involve transitions between potential wells. Molecular reaction dynamics has been extensively studied by means of crossed molecular beams, and a major application (and test) of potential energy functions for chemical reactions has been the calculation of vibrational states and angular distributions resulting from reactive scattering.

With high-quality beams of simple molecules, one can observe quantum mechanical oscillations in the angular distribution of scattered trajectories. These result from interference among alternative collision paths that yield indistinguishable outgoing molecular trajectories and states. A broader distribution of initial energies and angles will obliterate these fine features, however, yielding a smooth distribution of product trajectories.

In practice, calculations of molecular reaction dynamics commonly are based on the *quasiclassical approximation*. In this approximation, initial molecular states of vibration and rotation are described classically, but are chosen with initial energies and angular momenta that match quantum constraints. Quantum mechanical uncertainties in position are modeled by calculating many trajectories with randomly chosen vibrational and rotational phase angles. Trajectories are then computed by integrating the classical equations of motion. Finally, quantization of outcomes is modeled by lumping final trajectories into bins in phase space such that each bin corresponds to a permissible quantum state of the product. These calculations cannot yield quantum interference patterns or resonances, but give reasonably accurate descriptions of the coarse dynamical features of a reactive collision, such as the overall distribution of scattering angles and vibrational excitations.

Machine phase vs. molecular beams

Coarse features from a reaction dynamics perspective are, however, fine features from the perspective of a chemist or a nanomechanical engineer. Chemical reactions in nanomechanical systems bear little resemblance to reactions in crossed molecular beams. In an extended solid system subject to relatively slow mechanical motions, reactive molecular components do not encounter one another with well-defined energies and momenta; broad, thermal distributions dominate. Likewise, during the course of the encounter, the reactive components do no' form an isolated system with locally conserved energy, momentum, and angular momentum; the reactive system remains coupled to a thermal bath. The notion of a scattering angle is meaningless, and reaction-induced vibrational excitations are quickly thermalized.

In these respects, reactions in nanomechanical systems resemble the solution-phase reactions familiar to chemists. In such systems, the chief concern is with overall reaction rates, not with the details of trajectories. As a consequence, the detailed shape of the PES (crucial to the details of reactive scattering) is of reduced importance. Reaction rate data only weakly constrain PES properties. Given a few properties of the PES, thermally activated reaction rates are calculated using classical or quantum mechanical transition state theories based on statistical mechanics.

Generalized Trajectories

Classical dynamical models

Even in the absence of reactive transformations, the motions of typical nanomechanical systems cannot be completely described in terms of vibrations. Like protein chains or molecules in a liquid, these systems typically permit large displacements subject to complex, interacting constraints. Their motions must be described by more general methods.

It is common practice to model the trajectories of such systems by integrating the classical equations of motion, deriving the forces on each atom from approximate potential functions (e.g., molecular mechanics potentials). Where statistical properties are desired, a common approach is to integrate the equations of motion for a substantial time, taking a time average of the quantities of interest. Time steps are typically $\sim 10^{-15}$ s, and current computers have been used to follow the dynamics of $\sim 10^{3}$ atom systems for $\sim 10^{-9}$ s. These techniques are suitable for describing the short-term dynamics of nanomechanical components.

Quantum corrections

Quantum effects on molecular trajectories can be approximated by an adaptation of classical molecular mechanics techniques in which each atom is represented by a circular chain of atoms with suitable interactions; this has been applied to large molecules, such as the protein ferrocytochrome *c*. The effects are, as one would expect, largest for high-frequency vibrations, such as bond stretching, bending, and torsional motions involving hydrogen atoms. From a statistical perspective, thermal motions result in a certain probability density function for the positions of atoms with respect to their surroundings, and quantum effects broaden this probability density function.

Dynamical regimes

The dynamical behaviour of nanomechanical systems can be partitioned into three categories: (1) thermally excited vibrational motions within potential wells, (2) thermally excited transitions between potential wells, and (3) other motions, driven by nonthermal energy sources. Motions in category 1 can be treated either classically or quantum mechanically. Most motions encountered in category 3 are of low frequency and hence little influenced by quantum effects at ordinary temperatures. They can usually be modeled by applying classical dynamics to molecular mechanics potentials, or (on somewhat larger scales) to bounded continuum models.

Statistical Mechanics

Statistical mechanics (also known as *statistical* or *molecular thermodynamics*) is most commonly applied to relate macroscopic thermodynamic properties to statistical descriptions of the behaviour of large numbers of molecules. En route to describing properties of bulk matter, statistical mechanics frequently describes probability distributions for underlying molecular variables, such as position and velocity. In the present context, it is these probabilistic descriptions of the behaviour of individual molecular objects that are of primary value.

Statistical mechanics can frequently provide estimates of the statistical behaviour of nanomechanical systems without the computational cost of running a detailed dynamical simulation for long periods of time. Today, it is expensive to do a simulation of 10^3 atoms for as long as 10^{-9} s, yet a nanomechanism that fails even once per millisecond may be unacceptable. Estimating failure rates by observing more than 10^6 expensive simulations would be impractical. Statistical mechanics, in contrast, can provide estimates for the frequencies of extremely rare events using efficient analytical methods.

Statistical mechanics is commonly used to calculate quantities such as pressure, entropy, and free energy based on averages taken over many molecules in thermal equilibrium. But can temperature, for example, be used to describe a single molecule? There is no fundamental difference between (1) the statistical distribution of a dynamical quantity computed for many similar molecules at a particular instant (which is the usual concern of statistical mechanics), (2) the statistical distribution computed for a single representative molecule over a long period of time, and (3) the probability distribution for a single molecule at a single time. Accordingly, the concepts of pressure, entropy, free energy, and the like can be used to reason about (for example) the expected mean efficiency of a single nanomachine in a single operational cycle.

Detailed Dynamics vs. Statistical Mechanics

By omitting dynamical details, statistical mechanics provides a simplification that can assist both calculation and understanding. In the operation of real nanomechanical systems, the initial conditions are never known with the precision assumed in classical dynamical models, and seldom with the precision assumed in quantum dynamical models. Instead, the motions and displacements resulting from thermal excitation are random variables subject to some distribution. Rather than introducing arbitrary assumptions, statistical mechanics takes these uncertainties as fundamental, yielding inherently probabilistic descriptions of system behaviour.

The nanomechanical engineer's task, then, is to devise systems in which all probable behaviours (or all but exceedingly *improbable* behaviours) are compatible with successful system operation. This typically requires designs that present low energy barriers to desired behaviours, and high energy barriers to undesired behaviours, in which the high energy barriers often are a consequence of the use of stiff components. Even if one were to assume classical, deterministic behaviour and nearly perfect information regarding initial conditions, a typical nanomechanical system would soon require a statistical description. Consider the trajectory of a particle rebounding from an atom: because atoms are not flat, a small perturbation typically causes a particle to rebound at a different angle, leading to a larger perturbation at its next point of impact. In a typical system, trajectories that are initially almost identical will rapidly diverge until they have no similarity. This divergence is characteristic of chaos. Further, real nanomechanical systems are in mechanical or radiative contact with an environment at some nonzero temperature, providing a constant source of unpredictable thermal excitations.

Basic Results in Equilibrium Statistical Mechanics

Statistical mechanics takes its simplest form for systems at thermodynamic equilibrium. Since this is often a good approximation for real systems, some basic results are worth summarizing.

Quantum statistical mechanics

In quantum statistical mechanics, it is convenient to consider a system that is in thermal equilibrium with a heat bath, yet is assumed to have a set of bath-independent quantum states i = 0, 1,2,.... In equilibrium statistical mechanics, a complete description of a system consists of a specification of the probability, *P(i)*, for each state *i*. This takes the form

$$P(j) = \frac{\exp[-E(j)/kT]}{\sum_{i=0}^{\infty} \exp[-E(i)/kT]} \quad ...(1)$$

where *E(j)* represents the energy of state *j*. The probability that the system is in state *i* is proportional to the Boltzmann factor, exp[-*E* (*state*)/*kT*], and all states of a given energy are thus equally probable.

A quantity of special importance is the denominator of the above expression,

$$q = \sum_{i=0}^{\infty} \exp[-E(i)/kT] \quad ...2$$

where q is a temperature-dependent pure number termed the *partition function* of the system (note that its magnitude depends on the choice of the zero of the energy scale). The partition function can be related to the variables of classical thermodynamics. The mean energy of the system is given by an expression involving a constant-volume derivative of the partition function

$$\overline{E} = kT^2(\partial \ln q / \partial T)_V \quad ...(3)$$

as is the °entropy

$$S = [\partial(kT \ln q)/\partial T]_V \quad ...(4)$$

The °Helmholtz free energy is

$$F = -kT \ln q = E - TS \quad ..,(5)$$

and the pressure is given by a constant-temperature derivative

$$p = -(\partial F / \partial V)_T = kT(\partial \ln q / \partial V)_T \quad ...(6)$$

Classical statistical mechanics

Paralleling the quantum case, in classical statistical mechanics it is common to consider a system that is in thermal contact with a heat bath, yet is treated as having a bath-independent energy function, *E*(*state*). For a mechanical system, a state is defined as a point in the phase space defined by the set of position coordinates $q_1, q_2, q_3, \ldots, q_{3N}$ (the *position*), and the corresponding momentum coordinates $p_1, p_2, p_3, \ldots, p_{3N}$ (the *momentum*), where N is the number of atoms. Here, a complete description consists of a specification of the probability density function (PDF) over phase space. The fundamental result is

$$f_{state}(state) = \frac{\exp[-E(state)/kT]}{\int_{p,q}\cdots\int \exp[-E(state)/kT]\, dp_1 dq_1 dp_2 dq_2 \cdots dp_{3N} dq_{3N}} \quad ...(7)$$

where the PDF, $f_{state}(state)$, is the probability of occupancy per unit volume of phase space associated with each point in that space.

The denominator of the above expression (times a factor demanded by the correspondence principle) defines the *classical partition function* for a solid

$$q_c = (2\pi\hbar)^{3N} \int_{p,q}\cdots\int \exp[-E/kT]\, dp_1 dq_1 dp_2 dq_2 \cdots dp_{3N} dq_{3N} \quad ...(8)$$

which (in the classical approximation) can be used as the value of the partition function in Eqs. (3) through (6).

In nanomechanical design, a common concern is the probability that a thermally excited system will be found in a particular configuration at a particular time. Molecular mechanical systems can usually be described in terms of motion on a single potential energy surface, and the total energy can be divided into potential energy and kinetic energy terms

$$E(state) = V(position) + T(momentum) \quad ...(9)$$

With this division, Eq. (7) can be factored, and by integrating over the momentum coordinates of the phase space, a PDF over the position coordinates alone (the classical equilibrium PDF in configuration space) can be obtained

$$f_{position}(position) = \frac{\exp[-V(position)/kT]}{\int_{q}\cdots\int \exp[-V(position)/kT]\, dq_1 dq_2 \cdots dq_{3N}} \quad ...(10)$$

Note that the probability density associated with a position in configuration space is (save for a normalization factor) dependent purely on its potential energy. The distribution of momenta and kinetic energies is independent of the configuration, and hence the mean kinetic energy of a classical system at thermal equilibrium remains unchanged in all configurations.

Classical statistical mechanics is frequently a useful approximation. At ordinary temperatures its significant failings occur chiefly in stiff, high-frequency modes; these usually have little effect on the dynamics of nanometer-scale diamondoid components (which are by these standards soft, slow, and ponderous). Positional PDFs describing various elementary nanomechanical systems, all examined, comparing the results of quantum and classical models; its results indicate the limits within which classical statistical mechanics yields accurate results regarding positional uncertainty in nanomechanical engineering systems.

Configuration-space Picture

Although it yields no new physical information, it can be helpful to regard a classical mechanical system containing *N* atoms as a single moving point in a configuration space of *3N* dimensions, in which each of the three Cartesian coordinates of each atom corresponds to one dimension. Adding a single, orthogonal, "vertical" dimension to represent potential energy yields a potential energy surface in configuration space. The configuration point can then be imagined as sliding over an undulating, frictionless surface subject to a gravitational force—depending on initial conditions and the shape of the surface, the point may oscillate in a potential well, move along a valley, pass from one well to another through a °col between peaks, and so forth.

To make this dynamical picture work out properly, the configuration-space coordinates corresponding to an atom must vary in proportion to the Cartesian space coordinates of the atom multiplied by $m^{-1/2}$, where *m* is the mass of the atom. The kinetic energy of the coordinate point is then an isotropic function, depending only on the square of the speed.

In configuration space, a linear, elastic system corresponds to a point moving in a single potential well. For a two-atom system (approximating the bond as a linear spring and neglecting rotational degrees of freedom), this is a simple parabola. For a non-colinear *N-atom* system, the potential well retains a parabolic form along any line through the equilibrium point, and the isopotential surfaces are concentric $3N - 6$ dimensional ellipsoids; each of the $3N - 6$ axes of an ellipsoid represents the line of motion of a normal mode. A nonlinear system might permit (for example) the interchange of two atoms given a sufficiently great thermal excitation; the corresponding region of the potential surface has two wells joined by a pass.

Probability gas in configuration space

For a classical system at thermal equilibrium with a heat bath, statistical mechanics asserts that the probability density of the configuration point is an inverse exponential function of the energy, here represented by the height. Thus, the probability density varies across the configuration-space landscape much as the atmospheric density

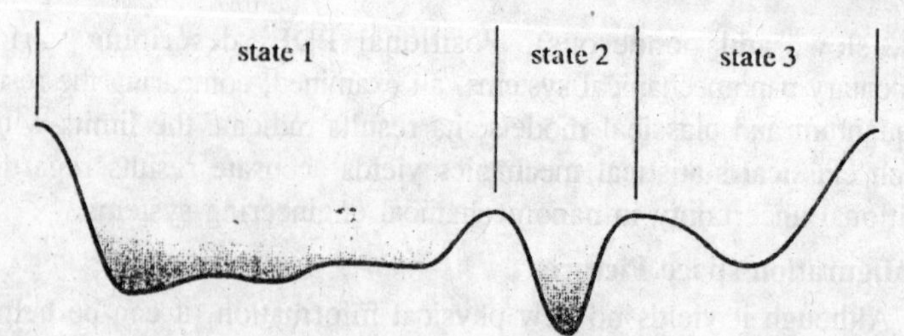

Fig. 8.1. A one-dimensional PES shaded to illustrate the probability gas visualization of the equilibrium distribution in configuration space.

varies across a real landscape. The configuration-space point is like a gas consisting of a single molecule, with a well-defined mean density, flux, and so forth, at every point.

As in an equilibrated atmosphere, the mean kinetic energy, and hence the temperature, is independent of the height of the land. The equilibrium ratio of the total probability in two connected valleys depends on their effective volumes; these depend on size and altitude, which correspond to the entropy and energy of the corresponding states. The rate at which probability diffuses through a col between the valleys depends on the height and width of the pass, and on the mean speed and overall probability density of the configuration point. All of these factors appear in transition state theory.

In statistical mechanics, the principle of *detailed balancing* asserts that, at equilibrium, the mean rate of transitions from state 1 to state 2 equals that from state 2 to state 1 for all pairs of states. For states defined as regions in configuration space, this has an intuitive interpretation. At equilibrium, gas molecules cross any arbitrarily defined surface element at equal rates from both sides; this is likewise true for the configuration-point gas, and for each surface element of the boundary separating any two states.

States in configuration space

The configuration space picture suggests one natural way to define what is meant by "*distinct states*" of a solid or liquid system. From each point on the energy landscape, there exists a path of steepest descent, and that path always ends in a point or region of locally minimal energy. Thus, points correspond to minima, and each local energy minimum can be taken to mark a distinct state.

Minima separated by barriers small compared to kT can often be regarded as a single minimum, because the barriers are crossed so easily. A system acting as a good rotary bearing, for example, will

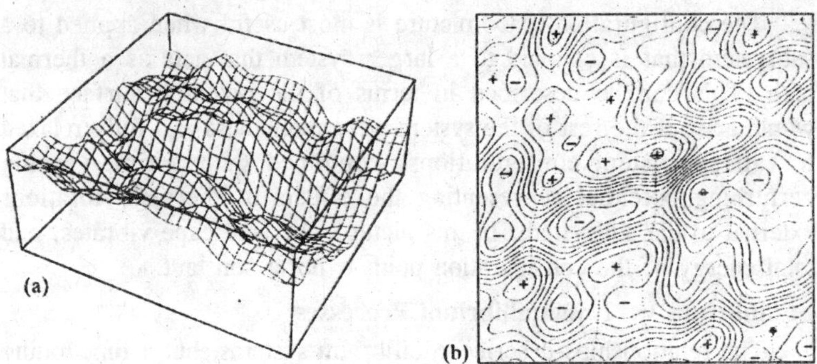

Fig. 8.2. Definition of states in terms of potential energy minima. Panel (a) shows a potential energy defined over a two-dimensional configuration space; Panel (b) shows the same surface as a contour map partitioned into regions corresponding to local potential energy minima.

exhibit a chain or loop of minima having essentially the same depth and separated by barriers of negligible height. For all practical purposes, a chain or loop of this sort constitutes a single potential well. In a properly designed nanomechanical system, cols between potential wells are either functional parts of the design, or are high enough to block any substantial probability flux.

Constraints and thermal excitation

Even without being able to visualize interconnected, approximately ellipsoidal potential wells, or ring-shaped valleys in a high-dimensional space, one can get a sense of the strongly constrained nature of the motion of the configuration point in such systems. A similar description of a chemically reacting solution-phase system lacks such well-defined features. Each possible covalent combination of atoms forms a separate valley, and the valleys intertwine in a manner that brings each into close proximity with many others. Which cols are available—determining which reactions can occur at appreciable rates—depends on the relative heights and widths of the numerous connections between valleys. A chemist attempting to control the pattern of reactions must exploit small differences in the heights and effective volumes (the energies and entropies) of cols and valleys.

In a nanomechanical system, in contrast, a reaction might occur between two groups brought together by a gear like rotary motion. The available configuration space consists of two ring-shaped valleys linked by a single col of modest height. Unwanted reactions can be prevented by gross mechanical constraints, rather than by small differences in energy and entropy.

The configuration-space picture is most useful when applied to a subsystem that is coupled to a larger system that acts as a thermal bath. This can be described in terms of the potential surface that would arise if the rest of the system were motionless and fully relaxed for all subsystem configurations, combined with a time-varying perturbing potential representing the effects of thermal vibrations external to the subsystem. In this picture, the landscape vibrates, and total energy of the configuration point is not a constant.

Equilibrium vs. Nonequilibrium Processes

The relationships describe equilibrium systems, but a functioning nanomechanical system, with motions driven by input power, is not at equilibrium. How useful, then, is equilibrium statistical mechanics?

Equilibrium statistical mechanics is seldom applied to a true equilibrium system. Matter under ordinary conditions of temperature and density, for example, has crystalline iron as its equilibrium state—if one allows true equilibration of all the nuclear degrees of freedom. In practice, the necessary reactions (e.g., fusion) proceed so slowly that they can be ignored. States (or dynamical domains) that are sufficiently °metastable can be treated as stable in calculating "equilibrium" properties.

In macromechanical systems, mechanical motion and thermal motion are quite distinct. In nanomechanical systems, this distinction usually remains clear. Thermal velocities may exceed the imposed mechanical velocities (which in most instances considered are quite small), but thermal velocities (unlike mechanical) have a mean value of zero and are independent of any driving force.

Deviations from equilibrium

In mechanical systems, frictional forces convert mechanical energy into °thermal energy in a spatially inhomogeneous fashion, causing thermal gradients. In estimating the fluctuations resulting from thermal motion, small regions of matter in such systems can then be approximated as being at thermal equilibrium save for deviations associated with temperature gradients, heat flows, and changing temperatures. When these deviations are small, equilibrium models give good estimates of the statistics of thermally induced displacements and motions in nanomechanical systems.

A small thermal gradient is one that produces only a small difference in the absolute temperature across the diameter of the system under analysis. For a nanomechanical system with a diameter of 10 nm, a 1% ΔT (at 300 K) requires a gradient of 3×10^8 K/m. Assuming a thermal conductivity of 10 W/m·K, this yields a thermal power flux of 3×10^9 W/m^2. This "small" thermal gradient produces a large (i.e., often unacceptable) temperature difference of 300 K in a modest 10^{-6} m distance.

A small thermal flux is one that produces only a small difference in the equilibrium distribution of thermal vibrations. This equilibrium distribution is characterized by large power fluxes (which cancel, at equilibrium) on the order of (*speed of sound*) × (*thermal energy*)/ (*volume*). For a typical material at ordinary temperatures, this is on the order of (3×10^8 J/m^3) × (10^3 m/s) = 3×10^{11} W/m^2. One percent of this value again corresponds to a net thermal power flux of 3×10^9 W/m^2. Since thicker layers result in unacceptable values of ΔT, a "small" power flux of this magnitude would be encountered in a working system only if a $\sim 10^{-6}$ m (or less) thick slab were to dissipate power at $\sim 10^{15}$ W/m^3 (or more). Despite the extraordinarily high *power-conversion* densities possible in small components, the *power-dissipation* densities for most complex, multicomponent systems will be small compared to this value.

A small rate of change of temperature is one that produces only a small ΔT during the characteristic vibrational relaxation time. In a crystal, a measure of this relaxation time is the phonon mean free path divided by the speed of sound; a typical value might be (10^{-8} m)/ (10^3 m/s) = 10^{-11} s. In a highly inhomogeneous medium, with highly anharmonic interactions, relaxation times typically are shorter. A 1% ΔT (at 300 K) during a single relaxation time then corresponds to a rate of temperature change in excess of 3×10^{11} K/s. For typical volumetric heat capacities ($\sim 10^6$ J/K·m^3), this "small" rate of temperature change requires a large power dissipation (as above, $\sim 10^{15}$ W/m^3 or more).

Process of thermalization

Because these "small" thermal gradients, thermal fluxes, and rates of temperature change are all so large, it is often acceptable to divide motions into mechanical and thermal components, describing the thermal motions in terms of the local temperature and the relationships of equilibrium statistical mechanics. The chief exceptions are in descriptions of the processes that convert mechanical motion into thermal excitation, where non equilibrium vibrational motions are generated and then thermalized. A typical nonequilibrium event is a transition from one potential well to another that is forced to occur at a particular time by an input of mechanical energy. In the configuration-space picture, the configuration point representing the subsystem passes through a col owing to an imposed change in the shape of the PES, and is thus injected into the new potential well with an unusually high energy and a somewhat-predictable momentum.

Regarding the new well as nearly quadratic, the initial state can be viewed as one with a disequilibrium distribution of excitation of normal modes. Relaxation then involves two processes: a tendency toward equilibration of the distribution of modal excitation via anharmonic interactions and a tendency toward equilibration of the expected total energy via interactions with a heat bath. For a general anharmonic system in weak contact with a thermal bath, a short-term description of the motion would describe families of trajectories; an intermediate-term description would treat all states of equal energy as equally probable, but would take account of the slowly decaying excess energy; and the long-term description would be in terms of the statistics of fully equilibrated thermal motion.

Entropy and Information

Entropy and probability

The transition from a detailed dynamical description to a statistical description entails discarding information. State variables that are regarded as having definite values in the dynamical description are regarded as having probability distributions in the statistical description. In quantum statistical mechanics, distinct states can be enumerated, and the entropy, which can be stated as

$$S = -k \sum_{\text{states}} P(state) \ln[P(state)] \qquad \ldots(11)$$

is a weighted measure of the number of possible states. This expression yields the familiar result (the third law of thermodynamics) that a perfect crystal at absolute zero has zero entropy: the structure of the crystal is known and it is in the vibrational ground state with unit probability; $1 \times \ln(1) = 0$, hence $S = 0$. This result is equally true for any completely specified structure at absolute zero.

Information can increase free energy

Probability, however, depends on the information one possesses. If I flip a coin and peek at it, it may be heads with probability one for me, while remaining heads with probability 1/2 for you. This may suggest that the entropy of an irregular solid at absolute zero is positive if these irregularities are unknown, but zero if they are completely described by some algorithm or external record (the complete specification just alluded to). If so, then entropy is not a local property of a physical system.

Studies of entropy and information confirm this view. In concrete terms, a structure can be made to yield more free energy if it is matched against another polymer with a sequence known to be identical—that is, if one has an external record of the sequence. Information regarding the sequence increases the Helmholtz free energy of the polymer, Eq. (5), because it eliminates the possibility of many alternative states, and thus lowers S. The free energy extracted from a system can thus depend on the existence of a record, initially located in the external environment, that is brought in, read, and returned unchanged. This understanding of entropy, although well established in the literature, contradicts the simple perspective presented in most textbooks.

Entropy and computation

These issues are intimately related to questions regarding the theoretical energy requirements of computation. In the early 1960s, Landauer observed that compressing the logical state of a computer entails compressing its physical state; for example, erasing or overwriting a one-bit storage device with previously unknown contents entails reducing its possible states (one or zero) to a single state (e.g., zero). Such a transformation dissipates a quantity of free energy $F \geq \ln(2)kT$ per bit erased.

Theorists have examined devices such as Brownian computers with idealized structures (i.e., hard, perfectly rigid parts of arbitrary shape) but subject to realistic dynamics and thermodynamics, as well as computational devices inspired by the molecular machinery of biological systems, idealized, deterministic mechanical systems, and quantum mechanical models. These and related studies indicate that logically reversible computations—those where the *output* uniquely specifies the *input*—can be performed in a manner approaching thermodynamic reversibility (that is, the dissipation of free energy per logically reversible operation can be made arbitrarily close to zero). The literature and results in this field have been well reviewed. They provide an improved understanding of the second law of thermodynamics, and are of direct relevance both to mechanical nanocomputers and to many other nanomechanical systems.

Uncertainty in Nanomechanical Systems

In conventional mechanical systems, the positions and shapes of components are never completely known: manufacturing tolerances, measurement inaccuracies, and environmental vibrations ensure this. When systems work despite these uncertainties, it is because the uncertainties are kept within tolerable bounds, either by reducing uncertainty or by expanding the range of tolerance.

In nanomechanical systems, quantum and statistical mechanics place physical limits on the reduction of uncertainties. For a structure of a given mass and stiffness in equilibrium at a given temperature, positional uncertainties are fixed and irreducible. Depending on the design of the system, these uncertainties can cause errors at a rate ranging from negligibly low to unacceptably high.

Quantum and classical parallels

There are strong *qualitative* parallels between the uncertainties of quantum mechanics and those of classical statistical mechanics. In both quantum and statistical mechanics, one begins with a potential energy function, $V(r)$. In quantum mechanics, the Schrodinger equation ensures that a particle of a given energy has a nonzero (though often vanishingly small) probability of penetrating a potential barrier of any finite height and thickness, and of being found in any region of space. In classical statistical mechanics at finite temperature, Boltzmann's law yields the same qualitative result by assigning a nonzero (though often vanishingly small) probability to sates of arbitrarily high energy. In quantum mechanics, a harmonic oscillator has a positional uncertainty characterized by a Gaussian probability distribution; classical mechanics at a finite temperature yields a result of the same form.

Quantum uncertainties measured as the product of the uncertainties in conjugate variables have an irreducible minimum, for example,

$$\Delta x \Delta p \geq \hbar / 2 \qquad ...(12)$$

but the uncertainty in either variable can be reduced to an arbitrary degree by a suitable measurement. Classical systems permit similar measurements, but present the illusion that the other variable can simultaneously be specified with arbitrary accuracy as well. This difference in the reducibility of uncertainty is, however, irrelevant in the context of equilibrium statistical mechanics. A measurement that reduces uncertainty also reduces entropy and hence alters the equilibrium of the system *from the perspective of the observer,* even if in the absence of a physical disturbance of the system itself. Thus, within the equilibrium statistical-mechanical description, uncertainties are irreducible by definition.

In a nanomechanical system, each component has many vibrational degrees of freedom, each subject to thermal excitation. Any attempt to use nanomechanical components within a system to represent the results of measurements performed on other nanomechanical components within a system can succeed in encoding only a small fraction of the information needed to represent the total vibrational state of the system. While one can imagine a device that uses measurement to reduce the uncertainties associated with one or a few critical components, systems will in general be dominated by components subject to statistical uncertainties that are in practice irreducible and that will usually be well described by equilibrium statistical mechanics.

Mean-force Potentials

Consider a piston that controls the volume of a cylinder containing an ideal gas. The work done in compressing the gas under isothermal conditions causes no change in the potential energy of the gas or of the piston, and yet so long as the system remains isothermal, the force exerted by the gas on the piston can be treated as the gradient of a potential energy function. This function defines a *mean-force potential*, since it describes the time-average forces experienced by the piston.

The use of a mean-force potential enables the statistical behaviour of a portion of an equilibrium (or near-equilibrium) system to be described without including all the coordinates of the full PES. In the piston-and-gas example, the use of a mean-force potential enables the statistical distribution of piston positions and velocities to be calculated directly, without reference to the motions of each of the gas molecules. In general, the mean-force potential defined over the retained coordinates is affected by the omitted coordinates to the extent that the position of the system with respect to the retained coordinates alters the permitted *range of motion* in the omitted coordinates. In the piston-and-gas example, the range of motion in the coordinates corresponding to the gas molecules is directly determined by the position of the piston; softer constraints on motion (e.g., time-varying °elastic restoring forces) have analogous effects. The use of a mean-force potential treats work of compression, which reduces entropy, as equivalent to work done against elastic forces, which increases potential energy.

In quantitative terms, the work done against a mean-force potential (in the classical model) is just $\Delta V - T\Delta S$; this equals ΔF, given that under isothermal conditions the change in kinetic energy $\Delta T=0$. A mean-force potential can be applied where imposed motions are slow (permitting thermal equilibration), or where motions—however fast—

are spontaneous and thermal, and hence result from the equilibrium state itself.

PES Revisited: Accuracy Requirements

Both dynamical and statistical mechanical models of molecular behaviour rely on potential energy surfaces that are (in all cases of nanomechanical interest) approximations known to deviate from reality. The scientific literature on potential energy surfaces describes efforts to improve the correspondence between experiment and theory, and hence focuses on the imperfections and limitations of existing models. In order to understand the utility of existing models from an engineering perspective, it is useful to consider the sensitivity of different physical phenomena to the existing inaccuracies.

Physical Accuracy

In chemical physics, experiments are designed to provide stringent tests of theoretical models of molecular systems (including their potential energy surfaces), and the theoretical models attempt to predict everything that can be experimentally observed. Physicists have made extensive use of molecular beam experiments in which molecules are prepared with precise momenta (and sometimes with control of vibrational states, rotational states, and polarization); they are then allowed to scatter (sometimes with a reactive exchange of atoms) and outcomes are observed and analyzed in terms of scattering angle (etc.).

The quantum interference effects that can be observed in such experiments provide a delicate test of potential energy surface models. Scattering events that involve bond formation and cleavage can traverse energy barriers of over 100 maJ, yet the interference effects are sensitive to much smaller energy differences. A characteristic molecular collision time is $> 10^{-13}$ s (the time required to travel 10^{-10} m at 10^{3} m/s); changing the potential energy along one of the interfering paths by 1 maJ or less shifts the phase of that path by a radian and causes a substantial change in the interference pattern. Since different trajectories explore different parts of the configuration space, describing interference effects correctly can require accuracy of this magnitude across the entire dynamically accessible potential energy surface.

Chemical Accuracy

Predicting solution-phase reaction

In solution chemistry, a standard challenge is to predict chemical equilibria, absolute reaction rates, and the relative rates of competing reactions. Synthetic chemistry can be viewed as an engineering discipline aimed at constructing molecules. In this task, rates and equilibria are of central importance: if a reaction equilibrates several species, then the yield of the desired product will depend on the equilibrium concentration ratios; alternatively, if a reaction can proceed along any of several effectively irreversible paths, then the yield of the desired product will depend on the ratio of the reaction rates. In some reactions, yields are affected by both rates and equilibria.

If entropic factors are equal, then the equilibrium ratio of two species is an exponential function of the difference in potential energy between the species. Likewise, if entropic factors (and certain dynamical factors) are equal, then the ratio of the rates of two competing reactions is an exponential function of the difference in potential energy between the two transition states. At 300 K, a difference of 1 maJ changes rates and equilibria by a factor of 1.27, a 10 maJ difference results in a factor of 11, and a 100 maJ difference results in a factor of 3.1×10^{10}. To a practicing chemist, an energy difference of 10 maJ between two competing species or transition states which changes the yield of a reaction from 8% to 90% usually matters more than would a 100 maJ shift in all energies (causing no change in the course of the reaction) or a 100 maJ shift in the transition-state energies which slows the reaction-completion time from a microsecond to an hour. In discussions of molecular energies, the term "chemical accuracy" usually implies errors of somewhat less than 10 maJ per molecule in describing the energies of chemical species (corresponding to potential wells) and transition states (corresponding to features of cols). Aside from en tropic effects dependent on the breadth of potential wells and cols, reaction rates typically exhibit only modest sensitivity to the shape—as distinct from the well depths and °barrier heights—of the potential energy surface.

Predicting the geometries of flexible structures

Potential energy functions are also used to predict molecular structures; the MM2 PES has good success for a wide range of small organic molecules. A more challenging test is protein modeling, where the shape and stability of the folded protein molecule depend on a delicate balance of free energy terms in which van der Waals interactions, overlap interactions, hydrogen bonding, torsional strains, and entropic factors all play crucial roles. The net stability of a folded structure typically is ~50 to 100 maJ, or ~0.01 maJ per atom. Although present molecular mechanics potentials are good enough to have found extensive use in protein modeling and design, their errors are large compared to the free energy of folding. Further, energy minimization usually yields structures that differ substantially from those experimentally determined by x-ray diffraction, even when the experimental structures are taken as a starting point.

Even for small organic molecules, a slightly inaccurate molecular mechanics model can predict structures that are totally wrong. Most molecules of concern in organic chemistry can exist in any of a number of conformations, differing by rotations about bonds; a simple example is *n*-octane. The relative energies of molecular conformations are sensitive to torsional energies and to weak, nonbonded interactions. Conformational equilibria, like other chemical equilibria, are greatly altered by energy differences of ~10 maJ. A crystal structure (a common source of geometric data) represents one of many possible molecular packing arrangements, again sensitive to weak forces. A small modeling error can result in a predicted crystal structure containing the wrong conformation packed in a lattice of the wrong symmetry.

Accurate Nanomechanical Design

Nanomechanical systems of the sort considered here are organic structures that resemble (or exceed) proteins in size, and some are used to perform chemical reactions. It is thus important to consider the sensitivity of nanomechanical designs to errors in potential energy surfaces.

Sensitive and insensitive designs

It will be possible to design nanomechanical systems that are exquisitely sensitive to the properties of a potential energy surface. For example, successful operation might be made to depend on a ratio of competing transition rates, as in conventional chemistry, or even on interference phenomena in angle-resolved scattering in crossed molecular beams. In general, any measurable physical property can be made essential to the correct operation of a suitable Rube Goldberg device, and hence any known class of unpredictable discrepancies between model and experiment can be used to design a class of devices that cannot with confidence be predicted to work. The design of such sensitive devices is closely related to good design practice in instrumentation and scientific experimentation, but is the opposite of good design practice in conventional engineering.

The robustness, predictability, durability, and performance of nanomechanical designs is usually maximized by making them from strong, stiff materials. Those considered here resemble diamond, consisting of highly °polycyclic, three dimensional networks of atoms linked by covalent bonds. Predictions of the stability and geometry of these diamondoid structures are far less sensitive to small errors in potential energy surfaces than are similar predictions for folded protein structures or conformationally mobile organic molecules.

"Perturbing" modeling errors

Increasing structural stiffness tends to reduce errors in predicted geometry caused by inaccuracies in the PES. Consider a multi atom design that, when modeled with a potential V_{model}, is predicted to have desirable geometrical properties. This function, however, differs from the correct (but unknown) potential, $V_{correct}$. The transition from model to reality may now be regarded as adding to the model system the perturbing forces defined by the gradient of the difference potential, $V_{diff} = V_{model} - V_{correct}$. A structure of greater rigidity suffers smaller deformations when subject to these perturbing forces; accordingly, a more rigid molecular design is more likely to survive the transition from model to reality without harmful distortion.

Good design practice can increase tolerance for the errors that remain. Nonetheless, errors in predicting molecular geometry can lead to designs that fail to work, in part because small local discrepancies in the predicted equilibrium geometry can have cumulative effects across large structures.

Modeling errors

Errors in modeling chemical reactions playa different role. If the correct sequence of reactions is performed, then the correct structure will result, regardless of small errors in describing reaction rates or transition-state geometries during the process. Distinct structures correspond to distinct states on a PES determined by invariant physical constants, and the manner of construction has no effect on the nature of the product. In conventional manufacturing, small variations in fabrication steps can have cumulative effects on properties such as product geometry, but no parallel problem arises in molecular manufacturing. Like switching events in digital logic, reactions either occur or they don't, and the result is either correct, or it isn't: ambiguous situations are unstable and transient.

In comparison to conventional chemical synthesis, molecular manufacturing processes based on positional control of synthetic reactions are less sensitive to small energy differences. Individual reaction steps can be driven to completion either by ensuring that energy differences greatly favour the product state, or by repeating trials until a molecular measurement verifies successful completion. To achieve regio- and stereospecificity, reactions can be guided not by differences in reaction rates and equilibria as in conventional synthesis, but by rigid control of reagent access to different sites. In chemical terms, this can create enormous ratios of effective reagent concentration between molecular sites that might otherwise be equally reactive. With positional control of synthesis, the main issues are achieving a high enough reaction rate (fast reactions best exploit the speed of nanomechanical systems) and avoiding (or being able to reverse) unwanted reactions immediately adjacent to the target site.

Testing of designs

At present, nanomechanical design must rely on modeling because the tools necessary for construction and testing are unavailable. Note, however, that errors in specific designs (as distinct from systematic errors regarding what sorts of designs are workable) will be of little significance until the systems they describe can be constructed and used. Then, of course, testing will become possible, and errors can be corrected.

CONCLUSIONS

Models of molecular dynamics (within the Born-Oppenheimer approximation) begin with a potential energy surface, but differ in their treatment of it. Nonstatistical classical methods proceed by assuming a set of initial conditions and then integrating the classical equations of motion of each atom, subject to the forces implied by the PES; statistical results describing (for example) a thermal distribution can be accumulated by examining long simulations. Statistical mechanics methods compute a statistical distribution, commonly that associated with thermal equilibrium: quantum methods compute a distribution of quantum states; classical methods compute a distribution of positions and velocities. Many nanomechanical systems can be analyzed with good accuracy using relationships based on classical equilibrium statistical mechanics, sometimes with quantum corrections. These relationships can be used to define a mean-force potential (or free-energy potential), enabling systems near equilibrium to be described in terms of fewer coordinates. A consideration of PES accuracy requirements in light of dynamical principles and nanomechanical concerns shows that the mechanical stiffness of diamondoid structures and mechanochemical devices often greatly reduces their sensitivity to PES modeling errors, relative to more familiar molecular systems.

9

MECHANOSYNTHESIS OF NANOMOLECULES

The term *mechanochemistry* was coined by Ostwald decades ago to describe "a branch of chemistry dealing with the chemical and physico-chemical changes of substances of all states of aggregation due to the influence of mechanical energy"; it has more recently been used to describe both the conversion of mechanical energy to chemical energy in polymers and the conversion of chemical energy to mechanical energy in biological mo:ecular motors; proponents of the first use urge that processes of the second sort be termed "chemomechanical" in accord with the direction of energy flow. As used here, *mechanochemistry* refers to processes in which mechanical motions *control* chemical reactions, guiding the molecular encounter and often providing activation energy; this is consistent with Ostwald's road definition. Since mechanochemical systems can approach thermodynamic reversibility, the direction of conversion between chemical and mechanical energy cannot be used as a defining characteristic. The present usage is thus related to but substantially distinct from those just cited. The term *mechanosynthesis* is here introduced to describe mechanochemical processes that perform steps in the construction of a complex molecular product.

Mechanochemistry, in the present sense, has features in common with reactions in organic crystals and in enzymatic active sites. In both crystals and enzymes, the environment surrounding a molecule can apply substantial stresses to it, or to a reaction intermediate, and can constrain reaction pathways so as to produce a specific product.

In crystals, these forces limit motion. Mechanochemical systems of the sort contemplated here, in contrast, characteristically transport reactive moieties over substantial distances, induce substantial displacements in the course of a reaction, and supplement local thermal and molecular potential energies with work done by external devices.

Mechanichemistry plays several roles in the systems. These include the transformation of small molecules, generation of mechanical power, preparation of bound reactive moieties, and mechanosynthesis of complex structures using such moieties. Although the discussion in the present chapter centers on reagent preparation and mechanosynthesis, the principles described are of general applicability.

Scope and Approach

Like solution-phase chemistry, machine-phase chemistry embraces an enormous range of possible reactions. Its potential breadth places the field beyond any hope of thorough exploration at present, and beyond any possibility of summarizing in the space of a chapter. Nonetheless, an attempt must be made to delineate the general capabilities and limitations of mechanochemical processes. How is this problem to be approached? In the absence of opportunities for immediate experimentation, two options suggest themselves:

The first approach relies on detailed theoretical modeling of mechanochemical processes. This yields quantitative descriptions of the sort familiar to those schooled in physics and engineering. Theoretical modeling suffers from several difficulties and limitations, however. Chemical reactions cannot be modeled using a molecular mechanics PES, but require either a specialized PES or the use of *ab initio* molecular orbital methods to calculate the energies of a set of molecular configurations. These quantum calculations have been expensive, and are of limited accuracy. Available techniques in computational quantum chemistry (to say nothing of future techniques) can be used to provide an extensive and detailed understanding of the capabilities of mechanosynthesis, but this will require considerable time, effort, and expense. Today, a survey of the capabilities of mechanosynthesis must rely on other methods.

The second approach begins by surveying the demonstrated capabilities of solution-phase organic synthesis, then examines how the conditions of mechanochemistry both add capabilities and impose constraints relative to this model.The resulting picture of the capabilities of mechanosynthesis thus is linked to experimental results in chemistry, rather than being directly founded on theory. This picture contains fewer quantitative descriptions than a physicist or an engineer may expect. If, however, we think that chemists are competent in chemistry as their achievements show), then we have reason to assume that their methodologies and modes of explanation are well suited to their subject matter. Organic synthesis is a notoriously qualitative field [it is "very little dependent on mathematical calculations," taking organic synthesis as its point of departure, accordingly represents a substantial departure from the style of exposition in the preceding chapters. Like much of chemistry, it relies heavily on example and analogy, reasoning from experimental results on model systems supplemented , by the general principles of chemical kinetics and thermodynamics.

PERSPECTIVES

Scale and Scope of Chemistry

The advent of organic synthesis commonly is dated from the 1828 synthesis of urea by Wohler, at a time when "chemists were just beginning to speculate about the arrangement of atoms within organic molecules". By the late 1800s, the basic concepts of atoms and molecular geometry were well established (except in the view of ardent Positivists), synthesis was subject to rational planning, and synthetic organic chemicals had become a major industry.

Today, a single nationally based professional group, the American Chemical Society, has ~150000 members. One year's output of *Chemical Abstracts* now occupies ~ 2 m of shelf space, containing abstracts of $\sim 5 \times 10^5$ articles; the volumes published to date occupy ~30 m of shelf space, with the indexes occupying ~30 m more. After 110 years of publication, *Beilsteins Handbuch der Organischen Chemie* occupies ~20 m of shelf space with over 300 volumes. The chemists' view of the current capabilities of organic synthesis is suggested by the observation that "the problem of synthesis design is simply that the number of possible synthetic routes to any target molecule of interest is enormous". This author estimates tens of millions of paths for a typical five-step synthesis, with vastly differing yields. A further indication is the barely cautious warning that "the idea persists that we can synthesize anything. The major flaw in this view is that it fails to recognize effectiveness and practicality".

During the first century of organic synthesis, chemists proceeded with no guidance from quantum mechanics, relying instead on a growing set of concepts regarding molecular geometry and thermodynamics, together with rules of thumb for estimating the energy differences between similar molecular configurations and the rate differences between similar reactions. These concepts and rules grew from efforts to rationalize experimental results; they are still being extended and remain central to organic synthesis to this day. Many reactions are known to yield particular results; somewhat fewer can be described in mechanistic detail (i.e., which atoms go where as the system passes through its transition state); far fewer have been modeled at the level of potential energy surfaces and molecular dynamics.

Qualitative Results

Organic synthesis is fundamentally qualitative, in much the same sense that topology (as distinct from, say, analytic geometry) is fundamentally qualitative. In synthesis, the chemist seeks to construct molecules with a particular pattern of bonding that can be described (give or take discrete, stereochemical differences) in topological terms. The discovery that substances of type A can be transformed into substances of type B is inherently qualitative and often of great value. The development of such useful species and processes as Grignard reagents, Wittig reactions, and Sharpless epoxidations are examples of contributions of this kind.

In describing an organic synthesis, numbers are used to specify temperatures, quantities of material, and characteristics such as melting points, refractive indexes, and so forth, but these numbers playa peripheral role. One can frequently examine several chapters of a monograph on chemistry while encountering no numbers save for dimensionless integers, such as those used to specify molecular structures and charge states. To be useful in experimental work, most measurements need only discriminate among different molecular species, and accurate discrimination need not require great precision. Many theoretical contributions of great fame and value take the form of nonquantitative rules (some subject to violation) that are expressed in terms of structure and geometry; examples include Markownikoff's rule for regioselectivity of addition to alkenes, the Alder rule for the production of isomers in Diels-Alder reactions, and the Woodward-Hoffmann rules regarding orbital symmetry in °cycloaddition processes.

Returning to the picture of potential energy surfaces in configuration space, the discovery of a new kind of reaction corresponds to the discovery of a set of similar cols between potential wells corresponding to similar molecular species. In standard chemistry, useful cols must be low enough to be thermally accessible at moderate temperatures and other cols must be relatively inaccessible (or harmless to traverse). The quantitative results of greatest interest in organic synthesis relate to the height and width of the cols (which determine reaction kinetics) and the height and width of the potential wells (which determine reaction equilibria).

Survey of Synthetic Achievements

No simple first-principles analysis leads from elementary facts regarding molecular physics to conclusions regarding the range of structures that can be made by diffusive, solution-phase chemical processes. Accordingly, one must examine the actual capabilities that chemists have developed. These include the synthesis of:

1. Highly strained molecules, such as cyclopropane rings with 60° bond angles, rather than the optimal 109.5°; cubane with a total strain energy of ~ 1.15 aJ, about twice the typical C-C bond energy; and [1.1.1]-propellane, in which the bonding configuration at two of the carbon atoms is inverted relative to the normal tetrahedral geometry-all four bonds are in one hemisphere.
2. Highly symmetrical molecules, such as dodecahedrane and cyclo[18]carbon.
3. Molecules that spontaneously decompose to free highly reactive species, such as carbenes and even free carbon atoms. (These species ordinarily exist only briefly before reacting.)
4. Some six hundred organic structures, containing a few tens to a few hundred atoms, identical to molecules synthesized by molecular machinery in living systems. Some examples include annotinine and endiandric acid C.
5. Chains of a hundred or more monomers, joined in a precise sequence to build biologically active protein and DNA molecules containing on the order of 1000 precisely bonded atoms.
6. Molecules designed to self-assemble into larger structures and even to catalyze the synthesis of copies of themselves.

Solution-phase Synthesis

Analytical Approach

This section uses the demonstrated mechanisms and capabilities of solutionphase organic synthesis as a basis for understanding the capabilities of mechanosynthesis. It proceeds by examining differences:

1. The solution-phase conditions and unit operations *excluded* by mechanosynthetic constraints.
2. The new conditions and unit operations *provided* by mechanosynthesis.
3. The additional system-level capabilities implied by these differences.

4. The additional constraints that must be satisfied to exploit these system-level capabilities.

Taken together with the known capabilities of diffusive organic synthesis, this set of difference gives substantial insight into the capabilities of mechanosynthesis..

Exothermic vs. exoergic

In solution-phase chemistry, when molecular potential energy is transformed into mechanical energy, this promptly appears as heat. Reactions that reduce potential energy are accordingly termed *exothermic*; those that increase it, *endothermic*. In mechanochemical systems, however, potential energy released by a reaction can often be stored elsewhere in the mechanical system, either as potential energy or as kinetic energy in the form of orderly motion. Accordingly, the more general terms *exoergic* and *endoergic* are appropriate. Changes in free energy determine the direction of a reaction, and since changes in entropy are usually small in mechanosynthetic steps (which typically transform one highly constrained structure into another), changes in free energy and in potential energy are often almost equal. Under these conditions, exoergic reactions can produce both heat and work, and most endoergic reactions are driven more by work than by heat.

Mechanochemical parameters

Two major themes in the following sections are reaction rates and reaction reliability in mechanochemical systems. Useful variables for an approximate analysis include the actuation time t_{act}, during which motions carry the system toward and then away from a reactive geometry; the transformation time t_{trans}, during which the geometry of the system permits a reaction to occur; the reaction rate k_{react}, which measures the probability per unit time of a reactive transformation during t_{trans}, given that a reaction has not yet occurred; the intersystem crossing rate k_{isc}, which measures the rate of the electronic intersystem crossing "reaction"; and the error rate k_{err}, which measures the rate of unwanted reactions. These parameters will be related to the probability of an error in a mechanochemical operation, P_{err}. It will usually be assumed that $t_{act} \approx 10^{-6}$ s, $t_{trans} \approx 10^{-7}$ s, and that each contribution to P_{err} is to be $\leq 10^{-15}$.

Basic Constraints

Loss of natural parallelism

Solution-phase synthesis often yields macroscopic quantities of product. In replacing a macroscopic volume of reagent solution with a single device, the number of reacting entities is reduced by a factor on the rough order of 10^{-23}. To compensate for this difference requires many mechanosynthetic devices, each operating at a high frequency.

Limitations on reagents and products

Mechanical control of reactions typically requires that reagent moieties be bound to extended "handle" structures; control of products typically requires that they, too, be bound, excluding the use or production of small, freely moving species. This constraint is less severe than it might seem, however, because free and bound species can often participate in similar reactions. For example, although an S_N2 reaction producing a free ion would cause unacceptable disorder, is permissible.

In reality, an ionic process like reaction would require a favourable electrostatic environment, and that environment could be used to bind the product ion. In water, small ions (from Li^+ to I^-) are found in solvent cages, and are bound with energies ≥ 330 maJ relative to vacuum. In condensed-phase acid reactions "H^+" is always bound. Reactions characteristic of small free radicals (e.g., H, F, OH) can frequently be made to proceed by concerted mechanisms in which the radicals are never free. Binding of such species usually lessens reactivity. Purely steric confinement sometimes is feasible, and has little effect on the electronic structure of the reagent. The requirement for handles (of any kind) places significant steric constraints on feasible reactions. Small, diffusing species can reach many sites (e.g., deep in pockets on a surface) that would be inaccessible to a mechanically positioned reagent moiety. In planning the synthesis of an object, geometric constraints are thus of increased importance. Finally, for reaction processes to be reliable, reagents must have adequate stability against unimolecular decomposition reactions.

Lack of true solvation

Fully eutactic systems must lack true solvation, since solvents are not eutactic. However, solvent-like eutactic environments can be superior to true solvents in promoting reaction speed and specificity. Mechanical transport can likewise substitute for diffusive transport; note that this sidesteps the constraint that reactants and products be soluble.

Limited usefulness of true photochemistry

Although similar results can be achieved using other sources of electronic excitation, photochemical reactions in a strict sense require photons. Optical wavelengths, however, are measured in hundreds of nanometers, and a cubic-wavelength block accordingly contains millions of cubic nanometers. This mismatch relative to the molecular scale limits the utility of photochemistry in mechanosynthesis.

Limitation to moderate temperatures

Mechanosynthetic systems are sensitive to damage from thermomechanical degradation, and the single-point failure assumption is adopted here. With careful design, reliable operation at substantially elevated temperatures may be feasible. Nonetheless, temperatures like those used in melt-processing of typical ceramics are almost surely infeasible. The present work assumes that temperatures remain near 300 K, thereby eliminating from consideration a large class of reactions that require elevated temperatures to proceed at an acceptable rate. This constraint could be partially evaded by exploiting highly localized, non thermal kinetic energy sources to provide activation energy (e.g., impacts driven by stored elastic energy), but this approach is not pursued here.

Basic Capabilities

Large effective concentrations

In a solution-phase reaction A + B → C, etc., the reaction frequency of molecules of type A is directly proportional to the concentration of molecules of type B (so long as molecules of B rarely interact with one another). In *bulk* chemistry, concentrations are commonly stated in moles per liter (M); here, they are stated as a molecular number density ($1\ nm^{-3} \approx 1.66$ M). The concentration of water *molecules* in liquid water is $\sim 33\ nm^{-3}$.

Because their moieties are held in proximity to one another, mechanosynthetic reactions resemble intramolecular reactions; these, in turn, often resemble intermolecular reactions, save for a covalent link between reacting moieties. From the rate for an intramolecular reaction (in s^{-1}) and the rate constant for an analogous intermolecular reaction (in $nm^3\ s^{-1}$), one can compute the *effective concentration* of a moiety B with respect to a moiety A in the intramolecular process by computing the (hypothetical) concentration of B-like molecules (in nm^{-3}) that would subject an A-like molecule to the same reaction frequency in the intermolecular process. The concept of effective concentration, although of no intrinsic importance to mechanosynthesis, is quite useful in estimating machine-phase reaction rates from solution-phase data.

Experiment shows that effective concentrations for reactions between moieties on small molecules are not limited to ~10 nm^{-3} (typical of immersion in a liquid); they can be as large as ~10^3 to 10^{10} nm^{-3}.

The occurrence of large magnitudes can readily be understood: In the probability-gas picture of classical transition state theory, the rate of transitions is proportional to the density of the gas. For concreteness and ease of visualization, consider an intramolecular reaction in which a single atom is transferred. The concentration of an atom at the bottom of the potential well defining its initial state is just the reciprocal of the effective volume; yields a number density of ~2×10^4 nm^{-3}; this local density is unrelated to (and far larger than) the mean number-density of nuclei in solids at ordinary pres5Ures. Further, many reactions require a relatively precise orientation of the reagent moieties; structural constraints in intramolecular reactions can greatly increase the concentration of properly oriented moieties, multiplying the effective concentration by a large factor.

Binding of reagents so as to ensure high effective concentrations is a major mechanism of enzymatic catalysis. Mechanochemical processes can likewise present reagent moieties to one another in favourable orientations and positions, producing high effective concentrations. Relative to intermolecular reactions in solution (with a typical concentration of ~1 nm^{-3}), this mechanism should routinely yield rate accelerations on the order of 10^4 or more.

Eutactic solvation

Where reactions involve substantial changes in charge separation, solvent effects can alter reaction rates by factors in excess of 10^5. Relative to vacuum or nonpolar environments, a polar solvent greatly decreases the energetic cost of creating dipoles or separated ions by orienting its molecules in a manner that decreases electrostatic field strengths. The feasible reduction in free energy is, however, lessened by the entropic cost of orienting the solvent molecules. A preorganized structure, in contrast, can equal the electrostatic effects of a good solvent without imposing entropic costs on the transition state.

Indeed, a pre organized environment can yield larger electrostatic effects. The solvent structure at the transition state of a solution-phase reaction minimizes the free energy of the transition-state system as a whole, including that of the solvent. A different organization of solvent molecules would minimize the free energy of the transition-state reactants alone; this organization typically is more highly ordered (imposing additional en tropic costs) and more highly polarized (imposing substantial energetic costs). Preorganized environments can be constructed to mimic this hypothetical, high-free-energy solvent state, paying the free-energy cost in advance and thereby outperforming a good solvent in promoting the desired reaction (a principle exploited by enzymes).

The magnitude of these effects can be roughly estimated. Ionic reactions commonly involve charge displacements on the order of one electron-charge and one bond length, or $\sim 2.5 \times 10^{-29}$ C·m. Fields of 2×10^9 V/m (e.g., 1V across 0.5 nm) are easily achieved in a polarized environment; contact potentials in metallic systems can produce fields of this magnitude. The energy difference associated with this field and charge displacement, ~50 maJ, gives a rough measure of the effect that local electrostatics can have on the energies of transition and product states in ionic reactions. This energy suffices to change rates and equilibria by factors of $>10^5$ at 300 K. The electrostatic energy difference between two orientations of an HCN molecule adjacent to a pair of oriented HCN molecules is ~110 maJ, an energy differential sufficient to change a rate or equilibrium by a factor $>10^{10}$.

Enzymatic active sites often resemble preorganized solvent structures, reducing the energy of a transition state. Mechanochemical systems based on a more general class of molecular structures can equal or exceed enzymes in this regard. The requirement that all structures be well bound, prohibiting true solvation, does not appear to be a significant sacrifice.

As enzymes show, molecular flexibility can be achieved without freely moving small molecules. In a mechanochemical context, however, it frequently is desirable to provide relatively rigid support for a reagent structure; one strategy is to surround the structure with a nonbonded (hence solvent-like) but rigid shell.

Mechanical forces

The term *piezochemistry* is in general use to describe solution-phase chemical processes in which mechanical pressure modifies chemical reactivity. Derived from the Greek *piezein* ("to press"), the term is here adopted to refer to a wider range of machine-phase chemical processes in which time-dependent mechanical forces (not necessarily a homogeneous, isotropic, slowly varying pressure) modify chemical reactivity.

In conventional piezochemistry, pressures accessible in commercially-available laboratory equipment (e.g., 0.1 to 2 GPa) frequently have substantial effects on reaction rates and equilibria. Transition states in solution-phase chemistry are characterized by an activation volume $\Delta V^{\ddagger}$ defined by Eyring in terms of a constant-temperature partial derivative:

$$\Delta V^{\ddagger} = -kT\left(\frac{\partial}{\partial p}\ln k_{\text{react}}\right)_T \qquad ...(1)$$

where k_{react} is the rate constant for the reaction and p is the pressure. Typical values of $\Delta V^{\ddagger}$ are in the range of –0.01 to –0.10 nm^3; values .:an be positive, for example, in fragmentation reactions. Many reactions have negative volumes of activation because their transition states combine bond making with bond breaking, and because the shortening of nonbonded distances by partial bond formation exceeds the lengthening of bonded distances by partial bond breakage. The relatively compact transition state is then favoured by increased pressure.

In the (poor) approximation that $\Delta V^{\ddagger}$ is independent of pressure, the rate increases exponentially with pressure. In reality, increasing pressures alter $\Delta V^{\ddagger}$ by altering the potential energy surface. Equation (1) and moderate-pressure values of $\Delta V^{\ddagger}$ give only a rough guide to the magnitudes of the effects that can be expected in higher-pressure piezochemical processes. With this assumption, a pressure of 2 GPa applied to a reacting system with $\Delta V^{\ddagger}$ = –0.02 nm^3 results in a rate increase of 1.6×10^4 at 300 K.

Nanomechanical mechanosynthetic devices can be built of diamond and diamondoid structures, and in the continuum approximation, stress is a scale-independent parameter. Accordingly, mechanosynthetic devices can apply pressures equaling those in macroscale diamond-anvil pressure cells. These reach ≥550 GPa, corresponding to a ≥30 nN compressive load per bond in a (111) diamond plane (neglecting the change in areal bond density with pressure). The effects of such pressures on bonding are substantial: at ~150 GPa, H_2 becomes metallic, as does CsI at ~110 GPa; xenon has likewise been metallized. Even a lesser pressure of 50 GPa (~3 nN/bond) can have large chemical effects. In the constant $\Delta V^{\ddagger}$ approximation, p = 50 GPa and $\Delta V^{\ddagger}$ = –0.01 nm^3 would yield a 500 maJ reduction in activation energy (and a physically unrealistic $> 10^{52}$ speedup); effects in real systems are smaller, but still large. The change in free energy resulting from high pressures can greatly exceed the change resulting from high temperatures.

Mechanically applied energy is subject to precise control. This discussion has been cast in terms of pressure and volume because it draws on experiments and theories applicable to solution-phase processes. In mechanochemistry, however, force and displacement are more useful variables, and (for a particular mode of displacement) an activation length

$$\Delta l^{\ddagger} = -kT\left(\frac{\partial}{\partial F}\ln k_{\text{react}}\right)_T \qquad \text{...(2)}$$

can be defined in terms of the applied force *F*. A typical magnitude for $\Delta l^{\ddagger}$ is ~ -0.1 nm, and feasible compressive loads can extend to >5 nN, which would once again yield a characteristic energy of ~500 maJ.

Mechanical instabilities can limit compressive loads: for example, if a single-atom tip is pressed against another atom, it has a tendency to slip sideways ("down off the hill") unless this is resisted by an adequate transverse stiffness. The negative stiffness associated with this instability has a magnitude

$$k_{s,\text{intab}} = -\frac{F_{\text{compr}}}{r_1 + r_2} \qquad \text{...(3)}$$

in the (conservative) model of hard spheres with radii r_1 and r_2. A typical transverse stiffness $k_{s\perp}$ for an unloaded tip is 20 N/m, and a typical atomic radius for a non bonded contact under substantial loads is ~0.1 nm, hence the condition

$$k_{s,\text{instab}} + k_{s\perp} > 0 \qquad \text{...(4)}$$

permits $F_{\text{compr}} \approx 4$ nN. A load of this magnitude suffices to store >100 maJ of potential energy in overlap repulsion between two unreactive atoms, and can store additional energy in the more-compliant deformation modes that lead toward a chemical reaction.

Unlike forces resulting from hydrostatic pressure, forces applied by mechanoochemical devices can be highly anisotropic and inhomogeneous on a molecular scale: large loads (including tension, shear, and torsion) can be applied to specific atoms and bonds in a controlled manner. These forces can cleave otherwise-stable bonds. Further, steric difficulties can be reduced by molecular compression and deformation in conventional piezochemistry. Under mechanochemical conditions, larger effects can be obtained which can significantly offset steric difficulties posed by the the mechanosynthetic requirement for bound reagents. In general, the availability of controlled forces of bond-breaking magnitude permits piezochemical modulation of reactions greatly exceeding that seen in solution-phase chemistry or in the comparatively low-strength, low-stiffness environment of an enzymatic active site.

Photochemistry

Mechanochemical system can exploit nonmechanical energy sources, for example, through electroemistry and energy transfer via electronic excitations. Both of these mechanism can be controlled more precisely in a mechanochemical environment than in solution or solution-surface systems.

Electrochemistry finds significant use in organic synthesis. Electrostatic potentials and tunneling rates can vary sharply on a molecular scale, permitting molecular-scale localization of electrochemical activity. Accordingly, electrochemical processes are well suited to exploitation in a mechanochemical context; they are also subject to modulation by piezochemical means. In electrochemical cells, pyridine can withstand an electrode potential of 3.3 V without reaction and tetrahydrofuran, -3.2 V, both with respect to a (catalytically active) platinum electrode; these potentials correspond to energy differences with a magnitude >500 maJ per unit charge. In field-ion microscopes (which provide one model for an electrode surface), electric fields can reach ~50 V/nm.

Direct photochemistry suffers from problems of localization, but photochemical effects can often be achieved by nonphotochemical means. Photochemical processes begin with the electronic excitation of a molecule by a photon, but this energy often can migrate from molecule to molecule as a discrete *exciton* before inducing a chemical reaction, and this process is highly sensitive to molecular structures and positions. Accordingly, the transfer of excitons can provide a better-controlled means for achieving photochemical ends; potential applications are, however, neglected in this volume.

Broadened options for catalysis

The structural requirements for mechanochemical reagents are satisfied by many catalytic structures: some are parts of solid surfaces already, and others (e.g., many homogeneous transition metal catalysts) have analogues that can be covalently anchored to a larger structure.

Aside from regeneration treatments (which are infrequent on a molecular time scale), conventional catalysts operate under steady-state conditions. In a typical catalytic cycle, reagents are bound to form a complex, the complex rearranges, and a product departs, all in the same medium at constant pressure, temperature, and so forth. If any transition state in this sequence of steps is too high in energy, its inaccessibility will block the reaction. If any intermediate state is too low in energy, its stability will block the reaction. If any feasible alternative reaction (with any reagent or contaminant in the diffusing mixture) leads to a stable complex, the catalyst will be poisoned. Many successful catalysts have been developed, but the above conditions are stringent, requiring a delicately balanced energy profile across a sequence of steps.

The range of feasible catalytic processes is broadened by the opportunities for control in mechanochemical processes. The elementary reaction in a catalytic cycle can occur under distinct conditions, lessening the requirement for delicate compromises to avoid large energy barriers or wells. Mechanochemical catalysts can be subjected to manipulations that modulate bond and transition state energies, typically by many times kT_{300}. Finally, comprehensive control of the molecular environment enables the designer to prevent many unwanted reactions, permitting the use of more reactive species.

Avoidance of competing reactions

In diffusive synthesis, achieving 95% yield in each of a long series of steps would ordinarily be considered excellent. At the end of a 100-step process, however, the net product would be ~0.6%; and at the end of a 2000-step process, $\sim 10^{-43}$%. A million tons of starting reagents would then reliably yield zero molecules of the desired product. At equilibrium, a reaction with $\Delta F \leq -145$ maJ at 300 K leaves less than 10^{-15} of the starting molecules unreacted. An energy difference of this magnitude is not uncommon, and a series of reactions with this yield would permit over 10^{10} sequential steps with high overall yield, in the absence of side reactions.

The complexity of the structures that can be built up by diffusive synthesis is limited not by an inability to add molecular fragments to a structure, but by the difficulty of avoiding mistaken additions. This problem is substantial, even in 100-atom structures. Moreover, as structures grow larger and more complex, they tend to have increasing numbers of functional groups having similar or identical electronic and steric properties (on a local scale). Reliably directing a conventional reagent to a specific functional group becomes increasingly difficult, and ultimately impossible.

In a well-designed eutactic mechanochemical system, unplanned molecular encounters do not occur, and most unwanted reactions accordingly are precluded. One class of exceptions consists of reactions analogous to unimolecular fragmentation and rearrangement. The other class of exceptions consists of reactions that occur in place of desired reactions; these can be termed *misreactions*.

A typical mechanosynthetic step involves the mechanically guided approach of a reagent moiety to a target structure, followed by its reaction at a site on that structure. In general, unguided reactions would be possible at several alternative sites, each separated from the target site by some distance (properly, a distance in configuration space). At one extreme, the alternative sites are separated by a distance sufficient to make an unwanted encounter in the guided case energetically infeasible (e.g., requiring that the mechanical system either break a strong bond or undergo an elastic deformation with a large energy cost). At the other extreme, the potential energy surface is such that passage through a single transition state leads to a branching valley, and then to two distinct potential wells, only one of which corresponds to the desired product; in this circumstance, unwanted reactions would be unavoidable. In intermediate cases, transi:ion states leading to desired and undesired products are separated by intermediate distances, and the mechanical stiffness of the guiding mechanism imposes a significant energy cost on the unwanted transition state, relative to the unguided case.

Considering the approach of a reagent moiety to a target structure in threedimensional space, a reaction pathway can be characterized by the trajectory of some atom in the moiety (such as an atom that participates in the formation of a bond to the surface). Inspection of familiar chemical reactions concurs with elementary expectations in suggesting that reaction pathways leading to alternative products commonly are characterized by trajectories that differ by a bond length ($d_t \approx 0.15$ nm) or more at the competing transition states. This is not universally true, but it is the norm; accordingly, avoidance of the exceptional situations imposes only modest limitations on the available sequences of chemical reactions. Where reagents are selective, or have reactivities with a strong orientational dependence, alternative trajectories usually are separated by far greater distances.

A reasonable stiffness for the displacement of an atom in a mechanically guided reagent moiety (measured with respect to a reasonably rigid or well-supported target structure) is 20 N/m (comparable to the ~30 N/m transverse stiffness of a carbon atom with respect to an sp^3 carbon). Note :hat the mechanical constraints on a reactive moiety can in many instances include not only the stiffness of its covalent framework, but forces resulting from an unreactive, closely packed surrounding structure that substantially blocks motion in undesired directions. Using the sinusoidal worst-case decoupling model to estimate error rates with nonsinusoidal potentials, the condition $P_{err} \leq 10^{-15}$ requires that the elastic energy difference between the two locations be $\geq \sim 180$ maJ at 300 K. With $k_s = 20$ N/m, this condition is satisfied at transition-state separations $d_t \geq 0.135$ nm. Accordingly, with modest constraints on the chemistry of the reacting species, suppression of unwanted reactions by factors of better than 10^{-15} should be routinely achievable. With $k_s = 30$ N/m and $d_t \geq 0.15$ nm, this model yields $P_{err} \leq 10^{-27}$. (Although the differing ratios of accessibility between alternative transition points at different times during the approach complicate the situation, the essential conclusions remain unchanged.) Discriminating between the ends of the double bond in an alkene during radical addition. It concludes that error rates $\leq 10^{-15}$ are achievable with stiffnesses of ~ 10 N/m, or ≤ 6 N/m using a trajectory-biasing technique.

Reliable control

It might seem that the most useful reagents would be those (e.g., strong free radicals, carbenes) that can react with many other structures. In solution-phase chemistry, however, unwanted reactions are the chief limit to synthesis, and reagents are prized less for their reactivity than for their selectivity. The ideal reagent in solution-phase synthesis is inert in all but a few circumstances, and it need not react swiftly when it reacts at all.

In mechanosynthesis, however, *selectivity* based on the local steric and electronic properties of the reagents themselves can be replaced by nearly perfect *specificity* based on positional control of reagent moieties by a surrounding mechanical system. Accordingly, highly reactive reagents gain utility, including reagents that are highly unstable to intermolecular reactions among molecules of the same type (e.g., benzynes, reactive dienes, and other fratricidal molecules). The use of reagents with increased reactivity can increase reaction frequencies, adding to (or, more properly, multiplying together with) the influences of high effective concentration, well-designed eutactic "*solvation*," and applied force.

Molecular Manufacturing

The capabilities of mechanosynthetic processes can be gauged by comparing the limitations with the strengths, taking the capabilities of diffusive synthesis as a baseline. This comparison suggests that mechanosynthetic operations are (overall) more capable than diffusive operations. These capabilities must, however, be judged in light of their proposed applications. This section previews proposed molecular manufacturing systems and discusses their associated speed and reliability requirements.

Molecular manufacturing approaches

Mechanochemical devices are of interest in this volume chiefly as components of mechanosynthetic systems capable of building large, complex structures, including systems of molecular machinery. The necessary positioning and manipulation of reagent moieties can be achieved in any of several ways, but these can be roughly divided into *molecular mill* approaches and *molecular manipulator* approaches. Molecular mills perform simple repetitive motions; molecular manipulators enable complex, programmable motions. Mills are well suited to the task of preparing reagent moieties for manipulators, but can also produce final products directly. In all cases considered, mechanical transport mechanisms replace diffusion.

Reagent preparation vs. application

Product synthesis can use reagent moieties that have been prepared from other reagent moieties, starting ultimately with simple feedstock molecules. The preparative steps can occur in an environment significantly different from that of the final synthetic steps.

In *reagent preparation*, the entire surrounding structure can be tailored to facilitate the desired transformation. In this respect, the reaction environment can resemble that of an enzymatic active site, but with the option of exploiting a wider range of structures, more active reagents, piezochemical processes, and so forth. The freedom to tailor the entire reaction environment during reagent preparation is a consequence of the small size and consequent steric exposure of the structures being manipulated.

In the final synthetic step, however, one reacting surface must be a feasible intermediate stage in constructing the product, and a prepared reagent must be applied to that surface. Construction strategies can be chosen to facilitate the sequence of synthetic reactions, but the freedom to tailor the entire environment solely to facilitate a single reaction is not available. In these *reagent application* steps, highly reactive moieties are of increased utility and moieties compatible with supporting structures of low steric bulk are desirable. These remarks apply both to manipulator-based systems and to the reagent-application stages of a mill-style system engaged in direct synthesis of complex products.

Reaction cycle times

One measure of the productivity of a manufacturing system is the time t_{prod} required for it to make a quantity of product equaling its own mass. Mill-style systems are anticipated to contain $\sim 10^6$ atoms per processing unit, with each unit responsible for converting a stream of input molecules into a stream of reagent moieties which are incorporated into identical sites in a stream of product structures. Manipulator-style systems are anticipated to contain $\sim 10^8$ atoms per manipulator, with each unit responsible for performing a series of mechanosynthetic operations on a single product structure. Each processing unit and manipulator operates on products at some frequency f_{synth}. A set of devices will transfer a mean number of atoms n_{synth} per operation; in anyone operation, the number transferred may be positive or negative (e.g., in abstraction reaction to prepare a radical site on a workpiece), but a typical value is $n_{synth} \approx 1$. Accordingly, if $f_{synth} = 10^3$ Hz, $t_{prod} = 10^3$ s for a mill-based system and 10^5 s for a manipulator-based system; for $f_{synth} = 10^6$ Hz, the corresponding values of t_{prod} are 1 and 100 s respectively. Values of $f_{synth} \geq 10^3$ Hz are acceptable for many practical applications, and $\sim 10^6$ Hz is used as a reference value in the systems analysis.

The exemplar calculations of the following sections assume that the time available for a reaction $t_{trans} = 10^{-7}$ s. This is compatible with $t_{act} \approx 10^{-6}$ s, or with $f_{synth} \approx 10^6$ Hz. These times are all long compared to the characteristic time scale of molecular vibrations, $\sim 10^{-13}$ s, but require reactions with low energy barriers.

Constraints on misreaction rates

In molecular manufacturing, two basic classes of error are (1) those that damage one product structure (fabrication errors), and (2) those that damage the manufacturing mechanism (destructive errors). Overall reliability can be increased in each instance by dividing a system (product or manufacturing) into smaller modules and replacing those that fail. Where fabrication errors occur, the module being made can be discarded before being incorporated into a product system. Where destructive errors occur, the manufacturing module can be stopped and its function taken over by one or more backup modules. Simplicity of design favours the use of relatively large modules, which in turn requires good reliability. For simplicity, the systems analysis assumes that all fabrication errors are destructive, placing further demands on reliability. For many purposes, 10^8-atom modules can be considered large. Fabrication errors at a rate of $\leq 10^{-10}$ per reagent application operation, resulting in a fabrication failure rate of $\leq 10^{-2}$, can thus be considered low. If we demand that each module in a manufacturing system process $\geq 10^2$ times its own mass before failing (and assume $n_{synth} \approx 1$), then a destructive-error rate of $\leq 10^{-10}$ per reagent application operation is again acceptable. The following calculations, however, take as a requirement that both classes of error occur at rates $\leq 10^{-15}$ per mechanosynthetic operation. An error rate of 10^{-15} is compatible with the assumptions; higher error rates could be tolerated at the cost of increased design complexity.

Meeting constraints

It has already been discussed the conditions for keeping the probability of misreactions $\leq P_{err} = 10^{-15}$; a related but distinct problem is to ensure that the desired reaction occurs, with a probability of omission (failure to react) $\leq P_{err}$. As also noted, a reaction with $\Delta F \leq -145$ maJ at 300 K can proceed with an equilibrium probability of remaining in the starting state $< 10^{-15}$. For a strongly exoergic reaction characterized by a rate constant k_{react}, the probability of omission in a single trial falls to $< 10^{-15}$ when the available reaction time $t_{trans} \geq 35/k_{react}$. A one-dimensional model based on transition state theory together with the bounds just described, yields a bound on the allowable barrier height for the reaction

$$\Delta V^{\ddagger} \leq kT \ln\left(\frac{t_{react} f_{TST}}{-\ln(P_{err})}\right) \qquad ...(5)$$

where f_{TST} is the transition state theory frequency factor; for mechanochemical reactions with relatively rigid, well-aligned reagent moieties, a reasonable value is $f_{TST} \geq 10^{12}$ s^{-1}. At 300 K, with $t_{react} = 10^{-7}$ and $P_{err} = 10^{-15}$, $\Delta V^{\ddagger} \leq 33$ maJ is acceptable.

How does this compare to typical solution-phase reactions? In the laboratory, characteristic reaction times (the reciprocals of the reaction rates) vary widely, from $< 10^{-9}$ to $> 10^6$ s; a not unusual reaction time in organic synthesis is $\sim 10^3$ s at a reactant concentration of ~ 1 nm^{-3}. Relative to this, achieving a reliable reaction in $t_{trans} = 10^{-7}$ s requires a speedup of $\sim 3 \times 10^{11}$. Increased effective concentration owing to mechanical positioning can easily provide a speedup of $> 3 \times 10^4$. Achieving the remaining factor of 10^7 requires that the energy barrier be lowered by ~ 70 maJ. Earlier sections have shown that electrostatic effects in eutactic environments can exceed 100 maJ, and that (crudely estimated) piezochemical effects can exceed 500 maJ; shifts from less-active to more-active reagents can likewise have large effects. Achieving adequate reaction speeds does not appear to be a severe constraint. The duration of t_{trans} can be increased by $\sim 10^2$ without undue cost by mechanisms that prolong the encounter process.

Intersystem crossing from the singlet state to a low-lying °triplet can cause errors, including omitted reactions and misreactions. Singlet transition-state geometries often resemble triplet equilibrium geometries, with corresponding reductions in singlet-triplet energy gaps. To ensure reliability, it is sufficient to ensure either (1) that the singlet-triplet gap $\Delta V_{s,t}$ always exceeds 145 maJ [$\approx kT \ln(10^{15})$ at 300 K], or (2) that as the $\Delta V_{s,t}$ increases, it exceeds 145 maJ at some time t_0 (prior to the time t_1 at which the geometry is no longer suitable for correct bond formation), and that the integral of the intersystem crossing rate k_{isc} meets the condition that

$$\int_0^{t_1} k_{isc}(t)\,dt \geq \ln(P_{err}) \qquad \text{...(6)}$$

Meeting a somewhat weaker but more complex condition would also suffice.

Conditional repetition

The conditions were chosen to assure a reliable reaction in a single trial, but this is unnecessary. A *conditional repetition* process uses the outcome of a measurement process to determine whether an operation is repeated. For example, consider a reaction process in which the reaction rate and time ensure equilibrium, but in which $\Delta F = 0$ rather than -145 maJ. This process has a .5 probability of success in any single trial. If a series of trials is terminated whenever success is achieved, then the mean number of trials required to achieve success is 2, and after 50 trials the probability of failure by omission is $< 10^{-15}$. A similar process with an exoergicity of 25 maJ would (at 300 K) have a .9976 probability of success in a given trial, requiring a sequence of 6 conditional repetitions to achieve $P_{err} \leq 10^{-15}$, with the mean number of trials before success ≈ 1.002.

Conditional repetition can be viewed as a variation on the chemical strategy of driving a reaction to completion by steadily removing product from a reacting mixture. A simple and efficient approach to implementing conditional repetition in a molecular mill.

Requirements for reagent stability

To meet the reliability objectives, reagent instability must not cause errors at a rate greater than the rate of misreactions. If the mean time between reactions is $\leq 100 t_{act} = 10^{-4}$ s, and the frequency factor for the instability is $\leq 10^{13}$ Hz, then limiting the instability-induced error rate per reaction to $\leq 10^{-15}$ requires a barrier height $\geq$ 230 maJ.

Many potential rearrangement and dissociation reactions can be suppressed in a mechanochemical context by surrounding the reagent moiety with a suitably structured environment. For example, nonbonded interactions that block motions on the pathway to a rearrangement before the structure reaches an alternative, deeper potential well can suppress the instability. Solid cage structures that block the escape of a molecular fragment can likewise suppress dissociative instabilities.

Since each of the failure mechanisms discussed in this and the previous sections is exponentially dependent on energy parameters, exceeding the specified objectives yields large improvements in reliability. For example, with a barrier height $\geq$ 275 maJ the error rate per reaction is $\sim 10^{-20}$, and sets of $\sim 10^4$ reagent moieties have a mean time to failure (via this instability) of $\sim$ 10,000 years.

Summary of the Comparison

Solution-phase chemistry has enabled the synthesis of small molecules with an extraordinary range of structures, but has not yet succeeded in constructing large structures while maintaining eutactic control. The specific comparisons made support some general conclusions regarding the relative capabilities of mechanosynthesis:

Versatility of reactions

Relative to diffusive synthesis, mechanosynthesis imposes several significant constraints on the kinds of reagents that can be effectively employed. It requires that reagent moieties be bound, which can reduce reactivity and impose steric constraints. It requires that reactions be fast at room temperature, limiting the magnitude of acceptable activation energies. It requires that reagent moieties have substantial stability against unimolecular decomposition reactions, precluding the use of some reagents that are acceptable in the diffusive synthesis of small molecules.

Offsetting these limitations, however, are several advantages. Fundamentally, *mechanochemical processes permit the control of more degrees of freedom than do comparable solution-phase processes;* these degrees of freedom include molecular positions, orientations, force, and torques. As a consequence, highly reactive moieties can be guided with precision, enabling the exploitation of reagents that are too indiscriminate for widespread use in solution-phase synthesis. Bound reagent moieties are subject to mechanical manipulation, enabling piezochemical effects to speed reactions and overcome substantial steric barriers through localized compression and deformation of molecular structures. Finally, entirely new modes of reaction become available when reagent moieties can be subjected to forces of bond-breaking magnitude. Overall, these gains in versatility appear to exceed the losses, and hence the range of local structural features that can be constructed by mechanosynthesis should equal or exceed the range feasible with diffusive synthesis.

Specificity of reactions

In diffusive synthesis, most reactions entail substantial rates of misreaction, and the probability of a misreaction during any given step tends to increase with the size of the product structure. Experience suggests that the cumulative probability of error becomes intolerable for product structures of more than a few hundred to a few thousand atoms.

In mechanosynthesis, reliable exclusion of misreactions can be achieved given (1) a sufficient distance between alternative transition states, and (2) a sufficient mechanical stiffness resisting relative displacements of the reagent moieties. Distances on the order of a bond length combined with stiffnesses comparable to those of bond angle-bending yield $P_{err} < 10^{-15}$. Error rates (per step) are independent of the size and complexity of the product structure, given that the product is either stiff or well supported.

Synthetic capabilities

Within the constraints required to achieve reliable, specific reactions, the versatility of the set of chemical transformations available in mechanosynthesis can be expected to equal or exceed the versatility of the set of transformations available in solution-phase synthesis. This versatility is sufficient to suggest that most kinetically stable substructures will prove susceptible to construction. Transformations that (1) satisfy these reliability constraints, and (2) yield kinetically stable substructures can then be composed into long sequences that maintain eutactic control and yield product structures of 10^{10} or more atoms.

Reactive Species

This section examines several classes of reagents from a mechanosynthetic perspective. *Numerous classes are omitted, and those included are discussed only briefly.* Polycyclic, broadly diamondoid structures are the products of greatest interest in the present context, hence this discussion focuses on the formation of carbon-carbon bonds; much of what is said is applicable to analogous nitrogen- and oxygen-containing compounds. Highly reactive species are of particular interest. Many of the following species would be regarded as reaction intermediates (rather than reagents) in solution-phase chemistry.

Ionic Species

Although ions are important in many solution-phase chemical systems, they are uncommon in the gas phase at room temperature; the electrostatic energy of an ion in vacuum can reduce stability by ≥500 maJ, favouring neutralization. In millstyle reagent preparation, reactions can occur in an electrostatically tailored environment; ionic species then can be as stable as those in solution and desired transformations can be driven by local electric fields. Accordingly, the utility of ionic species in these environments is, if anything, enhanced. In manipulatorbased reagent application, however, it may frequently be desirable to expose reagent moieties on a tip moving through open space toward a product structure that is not tailored for favourable electrostatics. The electrostatic energy of ionic species is then high (making solution precedents inapplicable), and the utility of ionic species in manipulator-based operations may be relatively limited.

Rearrangements and neutralization

Ionic species vary in their susceptibility to unimolecular rearrangement. Carbonium ions, for example, are prone to 1,2-shifts with low barrier energies: A reagent moiety in which this process can occur will likely fail to meet the stability criterion for use in molecular manufacturing processes. If, however, each of the R-groups is part of an extended rigid system, this rearrangement is mechanically infeasible. Rearrangement could also be prevented by steric constraints from a surrounding matrix, or (since the rearrangement involves charge migration) by local electrostatics.

Carbanions, having filled orbitals, are somewhat less prone to rearrangement and can more readily meet stability criteria. Again, structural, steric, and electrostatic characteristics can in many instances be used to supress unwanted rearrangements.

Charge neutralization provides another, nonlocal failure mechanism. To avoid this requires a design discipline that takes account of the ionization energies and electron affinities of all sites within a reasonable tunneling distance (several nanometers) of the ionic site, ensuring that charge neutralization is energetically unfavourable by an ample margin (e.g., ≥ 145 maJ).

Unsaturated Hydrocarbons

In diffusive chemistry, unsaturated hydrocarbons (alkenes and alkynes) find extensive use in the construction of carbon frameworks. Their reactions characteristically redistribute electrons from relatively high-energy pi bonds to relatively low-energy sigma bonds, thereby increasing the number of covalent linkages in the system while reducing the bond order of existing linkages. This process has a strongly negative $\Delta V^{\ddagger}$ and ΔV_{react}, since the increase in inter-atomic separation resulting from a reduction in bond order is outweighed by the ease resulting from the conversion of a nonbonded to a bonded interaction. Accordingly, these reactions are subject to strong piezochemical effects. Typical examples are Diels-Alder reactions which have $\Delta V^{\ddagger}$ in the range of ~ -0.05 to -0.07 nm^3.

Unsaturated hydrocarbons undergo useful reactions with ionic species, and their reactions with other reagents are touched on in the following sections.

From the perspective of solution-phase chemistry, mechanosynthesis has a greater freedom to exploit reactions involving strained (and therefore more reactive) alkenes and alkynes. The cyclic and polycyclic frameworks necessary to enforce strain will be commonplace, hence use of strained species can be routine; their use is desirable and practical for reasons discussed already. Planar alkenes have minimal energy, but molecules as highly pyramidalized as cubene and as highly twisted as adamantene have been synthesized. The reduced bonding overlap in these species makes the energetic penalty for unpairing of the pi electrons small enough to permit them to engage in diradical-like reactions under mechanochemical conditions.

Alkynes are of lowest energy when linear, but such highly reactive species as benzyne, cyclopentyne, and acenaphthyne, have been synthesized. Structures have short lifetimes in solution owing to intermolecular reactions, but all have mechanically anchored analogues that are kinetically stable in eutactic environments but highly reactive in synthetic applications. Again, one pi bond is sufficiently weak to permit diradical-like reactivity under suitable conditions. Also of high energy are allenes, cumulenes, and polyynes. Cumulenes and polyynes are of particular interest in building diamondoid structures consisting predominantly of carbon; they bring no unnecessary atoms into the reaction.

Unsaturated hydrocarbons are prone to various rearrangements, subject to constraints of geometry, bond energy, and orbital symmetry. Those shown here are stable, however, and (as with carbonium ions) mechanical constraints from a surrounding structure can inhibit many rearrangements of otherwise-unstable structural moieties .

Carbon Radicals

Free radicals result when a covalent bond is broken in a manner that leaves one of the bonding electrons with each fragment. Radicals thus have an unpaired electron spin and a half-occupied orbital. Radicals can be stabilized by delocalization, for example in pi systems, but most are highly reactive.

Reactions

Among the characteristic reactions of radicals are *abstraction*, in which the radical encounters a molecule and removes an atom (e.g., hydrogen), leaving a radical site behind, *addition* to an unsaturated hydrocarbon (here, a reactive pyramidalized species), generating an adjacent radical site on the target structure, and *radical coupling*, the inverse of bond cleavage. Radical addition and coupling have significant values of $\Delta V^{\ddagger}$ (for addition, ~-0.025 nm^3) and ΔV_{react}. Abstraction reactions often have a smaller $\Delta V^{\ddagger}$, and no significant ΔV_{react}. Their susceptibility to piezochemical acceleration is analyzed.

Radical coupling and intersystem crossing

Electron spin complicates radical coupling and related reactions. Bond formation demands that opposed spins be paired (a singlet state), but two radicals may instead have aligned spins (a triplet state), placing them on a repulsive PES. Bond formation then requires an electronic transition (triplet → singlet intersystem crossing). As the radicals approach, the gap between the triplet and singlet state energies grows, but this *decreases* the rate of intersystem crossing. In delocalized systems, bond formation can occur without intersystem crossing, at the energetic cost of placing some other portion of the system into a triplet state. If intersystem crossing is required during the transformation time, however, then a condition must be met, but with t_1 representing the time by which bond formation must have occurred. The condition that $\Delta V_{s,t} \geq 145$ maJ imposes a significant constraint because k_{isc} varies inversely with the electronic energy difference ΔV_{isc}, which (in the absence of mechanical relaxation) would equal the difference in equilibrium energies $\Delta V_{s,t}$, and will frequently be of a similar magnitude.

Values of k_{isc} for radical pairs in close proximity vary widely, and intramolecular radical pairs (diradicals) provide a model. For 1,3 and 1,4 diradicals, k_{isc} has been estimated to be comparable that of the $S_1 \rightarrow T_1$ intersystem crossing in aromatic molecules, where k_{isc} commonly is between 10^6 and 10^8 s^{-1}. Experiments with 1,3 diradicals in cyclic hydrocarbon structures found more favourable values of k_{isc}, ranging from $\sim 10^7$ to $> 10^{10}$ s^{-1}; the differences were attributed to conformational effects on orbital orientation, confirming rules.

The presence of high-*Z* atoms relaxes the spin restrictions on intersystem crossing by increasing spin-orbit coupling. In an aromatic-molecule model, k_{isc} for the $S_1 \rightarrow T_1$ transition in naphthalene is increased by a factor of ~ 50 by changing the solvent from ethanol to propyl iodide, and bonded heavy atoms have a larger effect: k_{isc} increases from $\sim 6 \times 10^6$ s^{-1} in 1-fluoronaphthalene to $> 6 \times 10^9$ s^{-1} in 1-iodonaphthalene. The energy-gap condition is met at ordinary temperatures in this system: ΔV_{isc} for 1-iodonaphthalene is 209 maJ.

From these examples, it is reasonable to expect that, in the absence of special adverse circumstances, the inclusion of high-*Z* atoms bonded in close proximity to radical sites can be used to ensure values of k_{isc} $\geq 10^9$ in radical coupling processes where $\Delta V_{s,t} \geq 145$ maJ. This is consistent with $P_{err} < 10^{-15}$ and $t_{trans} < 10^{-7}$ s. Failure to achieve intersystem crossing rates of this magnitude would increase the required value of t_{trans} for a particular operation, but would have only a modest effect on processing rates in a system as a whole, and no effect on the set of feasible transformations.

Types of radicals

Carbon radicals can broadly be divided into pi radicals and sigma radicals, depending on the hybridization of the radical orbital. Of these, sigma radicals are higher in energy and hence more reactive; examples include radicals at pyramidalized Sp^3 carbon (e.g., the 1-adamantyl radical), aryl radicals, and the alkynyl radical.

The alkynyl radical forms the strongest bonds to hydrogen and has excellent steric properties. *Ab initio* calculations on the ethynyl radical (HCC), however, predict a low-energy electronic transition, $A^2\Pi \leftarrow X^2\Sigma$, with an energy of only ~ 40 maJ. If alkynyl radicals have a similar state at a similar energy, they will have a significant probability of being found in the wrong electronic state. Interconversion, however, requires no intersystem crossing and should be fast compared to typical values of t_{trans}.

Radical rearrangement

Radicals are much less prone to rearrangement than are carbonium ions. Intramolecular abstraction and addition reactions of radicals are common, where they are mechanically feasible, are almost unknown unless the migrating group is capable of substantial electron delocalization (e.g., an aryl group).

Carbenes

Carbenes are divalent carbon species, formally the result of breaking two covalent bonds. The two nonbonding electrons in a carbene can be in either a singlet or a triplet state; the unpaired electrons in the triplet species behave much like those in radicals. Some carbenes are ground-state singlets in which the singlet-triplet energy gap $\Delta V_{s,t} \geq$ 145 maJ, making the probability of occupancy of the triplet state $\leq 10^{-15}$ at 300 K. Singlet carbenes can react directly to form singlet ground-state molecules; to achieve analogous results with triplet carbenes requires intersystem crossing.

Carbene reactions

Carbenes are of broad synthetic utility. They can undergo addition to double bonds, yielding cyclopropanes insertion in to C-H bonds, and coupling. These reactions frequently proceed with energy barriers of $\leq$20 maJ; many have a barrier of zero. Coupling of a singlet carbene and a radical should likewise proceed with little or no barrier.

Singlet and triplet carbenes

The prototypical carbene is methylene, a ground-state triplet with $\Delta V_{s,t} = -63$ maJ. Adamantylidene is thought to be a ground-state triplet. Carbenes with better steric properties (i.e., with supporting structures occupying a smaller solid angle) tend to be ground-state singlets: reducing the bond angle at the carbene stabilizes the singlet state, as does a double bond. In cyclopropenylidene the predicted value of $\Delta V_{s,t}$ is $\sim$490 maJ; in vinylidene, a prototype for alkylidenecarbenes, $\Delta V_{s,t}$ $\approx$ 320 maJ; and in cumulenylidenecarbenes such as, $\Delta V_{s,t}$ >400 maJ. These gaps are consistent with reliable singlet behaviour in a molecular manufacturing context. Since singlet states of carbenes are appreciably more polar than triplets, singlet states can be significantly stabilized by a suitable electrostatic environment. For diphenylcarbene, experiment indicates that the shift from a nonpolar solvent to the highly polar acetonitrile increases $\Delta V_{s,t}$ by $\sim$10 maJ. A preorganized environment having a polarization greater than that induced in a solvent by the singlet dipole can increase the stabilization. Substituents including N, O, F, and Cl also tend to stabilize the singlet state.

Carbene rearrangements

Carbenes have a substantial tendency to rearrange; alkylcarbenes, for example, readily transform into alkenes. Among unsaturated carbenes, alkadienylidenecarbenes and cumulenylidenearbenes have no available local rearrangements at the carbene center, and although vinylidene itself readily transforms to ethyne, reaction is not observed, and related cyclic species.

Organometallic Reagents

Various reagents with metal-carbon bonds are used in organic synthesis; this section briefly discusses only a few classes.

Grignard and organolithium reagents

Grignard reagents (RMgX) and organolithium reagents (RLi) find extensive use, providing weakly bonded carbon atoms with a high electron density. As is common with organometallic species, their metal orbitals can accept electron pairs from coordinated molecules; this provides options for improving stability and mechanical manipulability. For example, in an ether solution a typical Grignard eagent structure is organometallic species, that might be used in mechanosynthesis.

Transition-metal complexes

Complexes containing a *d*-block transition-metal atom exhibit versatile chemistry; such complexes are prominent in catalysis, including reactions that make and break carbon-carbon bonds. The presence of accessible *d* orbitals in addition to the *p* orbitals available in first-row elements changes chemical interactions in several useful ways: the orbital-symmetry constraints of reactions among first-row elements are relaxed, and bonded structures can have six or more ligands (rather than four); further, the relatively long bonds (typical M-C lengths are ~0.19 to ~0.24 nm) reduce steric congestion, thereby facilitating multicomponent interactions. (Longer bonds result in coordination shells with areas ~1.6 to 2.6 times larger than those of first-row atoms). Many transition-metal complexes readily change their coordination number and oxidation state in the course of chemical reactions. Electronic differences among transition metals are large; complexation further increases their diversity. Clusters containing multiple metal atoms eventually resemble metal surfaces, which also find widespread use in catalysis.

Transition metals in bulky complexes appear more useful in reagent preparation and small-molecule processing than in sterically constrained reagent-application operations. In enfolded sites, ligand arrangements can be determined by mechanical constraints in the surrounding structure and placed under piezochemical control.

Forcible Mechanochemical Processes

General Considerations

Force and activation energy

Forces in piezochemical processes alter the reaction PES, reducing the activation energy; in some instances, they can eliminate energy barriers entirely, thereby merging initially distinct states. Barrier reductions of ~70 maJ can convert reactions that take 10^3 s in solution into reactions that complete reliably in 10^{-7} s. How much force is required to have such an effect?

Initial motion along a reaction coordinate can resemble the stretching of bonds or the compression of nonbonded contacts. As these motions continue, the resisting forces increase, but (usually) not so rapidly or so far as they would in simple bond cleavage or in the compression of an unreactive molecular substance. The potential energy curve instead levels off, passes through a transition state, and falls into another well. Since the energy stored in a given degree of freedom by a given force is proportional to compliance, the energy stored by a force applied through an unreactive bonded or nonbonded interaction is usually lower than that stored in a reactive system.

A compressive load of ~2 to 3 nN stores 70 maJ in a nonbonded interaction in the MM2 model; the ~30 nN per-bond compressive load in a diamond anvil cell is an order of magnitude larger. The energy stored in bonds is more variable, but for a C—C bond in the Morse model, a tensile force of ~5 nN stores 70 maJ.

Applied forces and energy dissipation

When actuation times are relatively long (~10^{-6} s), acoustic radiation from time-varying forces is minimal, as are free-energy losses resulting from potential-well compression, given reasonable values for critical stiffnesses. Likewise, with small displacements (~1 nm) and low speeds (~10^{-3} m/s), phonon-scattering losses are small. In an elementary reaction process, the most significant potential sources of dissipation are transitions among time-dependent wells.

Although the issue is distinct from the basic qualitative question of mechanosynthesis (i.e., what can be synthesized?), minimizing energy losses is of practical interest. Losses can broadly be divided into three classes: (1) those that are many times *kT*, resulting from the merging of an occupied high-energy well with an unoccupied low-energy well; (2) those on the order of kT, resulting from the merging of wells of similar energy; and (3) those of negligible magnitude, resulting from the merging of a low-energy, occupied well with a high-energy, unoccupied well. The simplest way to achieve high reaction reliability is to follow route (1), dissipating ≥145 maJ per operation. During forcible mechanochemical processes, however, it will in many instances be possible first to follow route (2) or (3) to a state in which the wells are merged (or rapidly equilibrating over a low barrier), then to use piezochemical effects to transform the PES to a type (1) surface before separation. This yields a process with reliability characteristic of (1), but with energy dissipation characteristic of (2) or (3). Systems capable of altering relative well depths by ≥150 maJ in mid-transformation can achieve error rates $< 10^{-15}$ with an energy dissipation $< 0.1\ kT_{300}$. Opportunities for this sort of control are discussed in several of the following sections.

Tensile Bond Cleavage

Cleavage of a bond by tensile stress is perhaps the simplest mechanochemical process, providing an instance of the conversion of mechanical energy to chemical energy and illustrating the relationship between stiffness and thermodynamic reversibility. Further, tensile bond cleavage plays a role in several of the mechanosynthetic processes described in later sections.

A typical C—C bond has a relatively large strength. k_s for such a bond is lower than that for bonds to several other first-row elements, but higher than that for bonds to second-row elements. In many practical applications, the bond to be cleaved is of lower strength and stiffness than a typical C—C bond. Cleavage of the C—C bond will be considered in some detail, however, and can serve as a basis for comparison to other bond cleavage processes.

Load and strength

The 300 K bond-lifetime curves indicate the tensile loads required to effect rapid bond cleavage. To achieve a level of reliability characterized by P_{err} requires a barrier meeting the condition; for a C—C bond in this model, achieving $P_{err} \leq 10^{-15}$ within $t_{trans} = 10^{-7}$ s requires a tensile load of ~4.2 nN. The Morse potential underestimates bond tensile strengths, but the problem of achieving sufficient tensile loads for rapid bond cleavage essentially pits the strength of one bond against that of others, hence errors in estimated strengths approximately cancel.

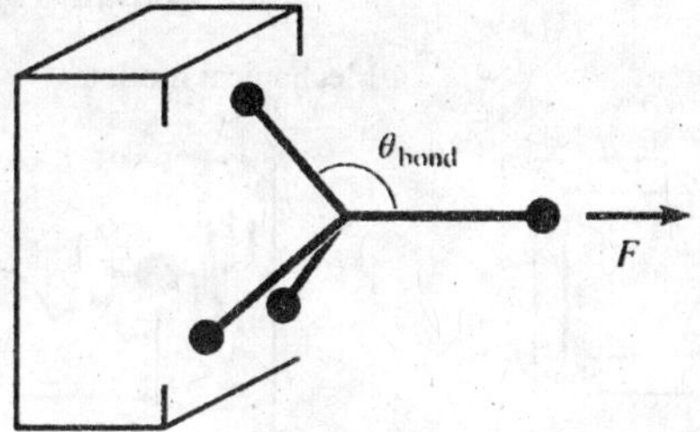

Fig. 9.1. Bond angle in a distorted tetrahedral geometry.

As long as a carbon atom occupies a site with tetrahedral symmetry, straining one bond to the theoretical zero-Kelvin, zero-tunneling breaking point necessarily does the same to the rest. To concentrate a larger load on one bond requires that the angle θ_{bond} be increased, thereby increasing the alignment of the back bonds with the axis of stress and reducing their loads. With load, however, even an initially tetrahedral geometry distorts in the desired fashion. Increase of θ_{bond} from 109.5° to 115° reduces back-bond tensile stresses to 0.79 of their undistorted-geometry values. Breaking of a back bond under these conditions would of necessity be a thermally activated process, and the energy barrier for breaking more than one bond at a time would be prohibitive. Moreover, in the structures considered here, breaking of a single bond is strongly resisted by angle-bending restoring forces from the remaining bonds (a ~0.1 nm, bond-breaking deformation is associated with an angle-strain energy of ~300 maJ); as has been discussed, this solid-cage effect invalidates the model and strongly stabilizes structures. These stress and energy differences are more than adequate to ensure a $> 10^{15}$ difference in rates of bond cleavage.

Stiffness requirements for low-dissipation cleavage

Transitions between time-dependent potential wells can cause energy dissipation, and distinct wells are of necessity separated by regions of negative stiffness in the potential energy surface. The potential energy for a pair of atoms undergoing bond cleavage can be described as the sum of the bond energy and the elastic deformation energy of the structures in which the atoms are embedded:

$$V_{\text{cleave}}(\Delta r, \Delta d) = V_{\text{bond}}(\Delta r) + \tfrac{1}{2} k_{\text{s,struct}} (\Delta d - \Delta r)^2$$

where Δd and Δr, and the function $V_{bond}(\Delta r)$ is the bond potential energy. The elastic deformation energy (neglecting modes orthogonal to the reaction coordinate) is a function of the displacement be-the bonded atoms and their equilibrium positions with respect to the supporting

Structural diagram: **Mechanical model:**

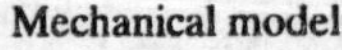

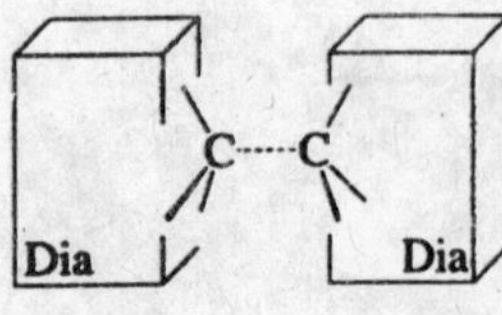

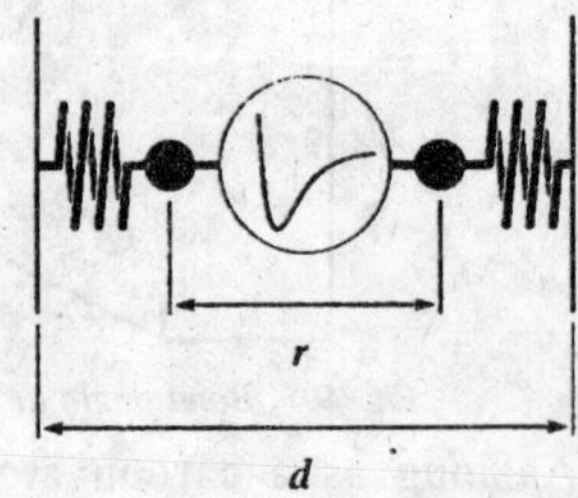

Fig. 9.2. Diagrams illustrating tensile bond cleavage and the corresponding coordinates.

structure, and corresponds to some positive stiffness $k_{s,struct}$. The bond function, however, has a negative stiffness in the bond-breaking separation range. The Morse potential predicts an extreme value of −0.125k_s, where k_s is the stretching stiffness of the bond at its equilibrium the Lippincott potential predicts negative stiffnesses of greater magnitude (−115 rather than −55 N/m, for a standard C–C bond), and hence is more conservative in the present context. It is adopted in the following analysis.

The potential energy curves illustrate as a function of Δr for various values of Δd for one set of model parameters. As can be seen, a steady increase in Δd causes the evolution of the system from a single well, to a pair of wells, to a single well again; larger values of $k_{s,struct}$ first reduce and then eliminate the barriers. At finite temperatures and modest speeds, transitions can occur without causing substantial

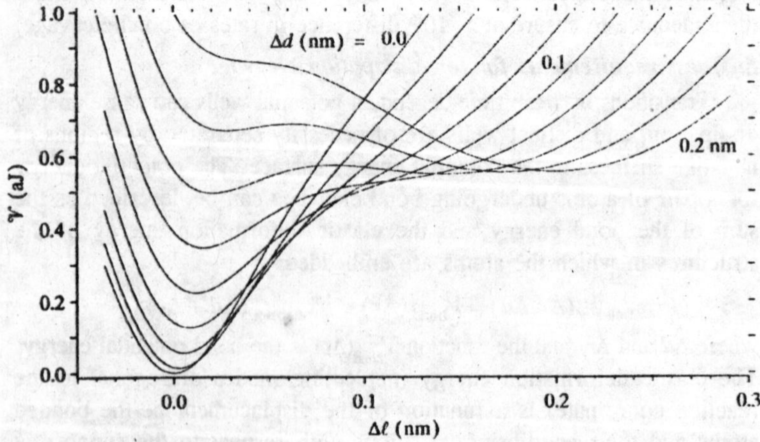

Fig. 9.3. Potential energy as a function of CC distance, for several values of support separation.

dissipation, so long as good equilibration occurs between the old well and the new before the energy of the new well has fallen substantially below that of the old. This requires that, during the time when the wells are of nearly equal energy, the mean interval between transitions be short compared to the time required for significant changes in relative well depth to occur. A transition rate of 10^9 s^{-1} ensures low dissipation (small compared to kT) when the characteristic time for the evolution of the wells is $\sim 10^{-7}$ s. Assuming a frequency factor of 10^{13} s^{-1}, this is achieved for barrier heights $\leq$38 maJ at 300 K.

For estimating energy dissipation, a conservative measure of the barrier height for the process as a whole is the height when the two wells are of equal depth, $\Delta V_{=}^{\ddagger}$ (assuming substantial values of $k_{s,struct}$ > to limit the entropic differences between the two wells). Fig. 9.4 plots $\Delta V_{=}^{\ddagger}$ as a function of $k_{s,struct}$ for several bond types. For processes in which $k_{s,struct}$ >90 N/m, and characteristic times are $\geq 10^{-7}$ s, dissipation is small relative to kT for all the single bonds shown; for standard C—C bonds, $k_{s,struct}$ >60 N/m is sufficient.

Spin, dissipation, and reversibility

In the absence of intersystem crossing, bond cleavage yields a singlet diradical. In a well-separated diradical, however, the singlet-triplet energy gap approaches zero, and thermal excitation soon populates the triplet state. If bond cleavage is fast compared to intersystem crossing, this equilibration process results in ΔF=-ln(2)kT (corresponding to the loss of one bit of information). Conversely, if intersystem crossing

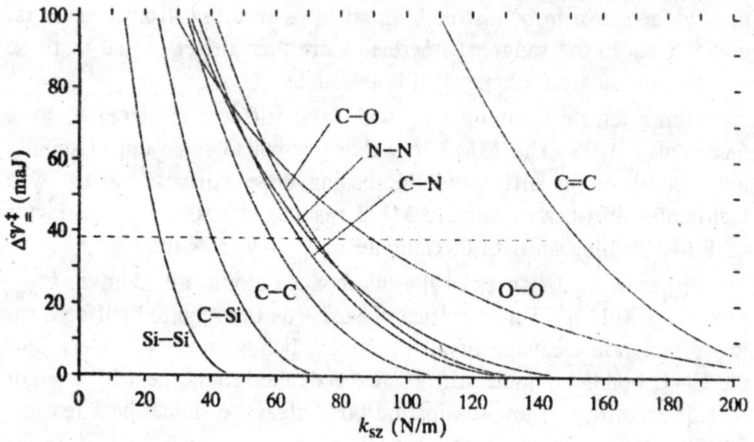

Fig. 9.4. Barrier heights vs. stiffness, for various bonds placed under a tensile load which equalizes the well depths.

is fast, the thermal population of the (repulsive) triplet state results in a reduction of the mean-force bond potential energy during cleavage, and no significant dissipation. Note that slow intersystem crossing can cause large energy dissipation in mechanically forced radical coupling, even when the reverse process has a dissipation $<kT$.

Atomic stiffness estimation

In the linear, continuum approximation, the *z*-axis deformation of a surface at a radius *r* from a *z*-axis point load is $\propto 1/r$. Accordingly, most of the compliance associated with displacement of an atom on a surface results from the compliance of the portion of the structure within a few bond radii.

A carbon atom on a hydrogenated diamond (111) surface can be taken as a model for sp^3 carbons on the surfaces of diamondoid structures. The MM2 model value for the *z*-axis stiffness k_{sz} of such a carbon atom on a semi-infinite lattice can be accurately approximated by measuring the stiffness in a series of approximately hemispherical, diamond-like clusters of increasing radius, holding the lattice-terminating hydrogen atoms fixed. In the continuum model, the compliance of the region outside a hemisphere is $\propto 1/r$; treating the number of carbon atoms as a measure of r^3 and fitting the four cluster-stiffness values with this model yields $k_{sz} = 153 \pm 2$ N/m.

This model holds for small displacements, but at the larger displacements corresponding to peak forces in the bond-cleavage process, bond angles and lengths are significantly distorted; this affects stiffness. Examination of the energy as a function of *z*-axis displacements shows that the decrease in bondstretching stiffness resulting from tension is, under loads in the range of interest, more than offset by the increase in k_{sz} resulting from changes in bond angle.

Under tensile loads of 3 to 7 nN, the stiffness is increased by a factor of ~ 1.05. The MM2 model is known to have unrealistically low bond-bending stiffnesses; increasing these stiffness values by a factor of 1.49 (to approximate MM3 results) increases k_{sz} by a factor of 1.14, yielding an overall estimate of $k_{sz} \approx 183$ N/m.

Since the compliances of the two bonded atoms are additive, $k_{s,struct} = k_{sz}/2 \approx 90$ N/m. This significantly exceeds the required stiffness for low-dissipation cleavage of all the single bonds, save for O—O. For the C—C bond, available stiffness exceeds the requirement by a factor of 1.5; accordingly, low-dissipation bond cleavage should be a feasible process in a broad range of circumstances.

Abstraction

Abstraction reactions can prepare radicals for use either as tools or as activated workpiece sites. Although various species are subject to abstraction, the present discussion focuses on hydrogen. Most exoergic or energetically neutral hydrogen abstraction reactions have activation energies ≤100 maJ. The abstraction reaction

$$H_3C\bullet + HCH_3 \rightarrow H_3CH_3 + \bullet CH_3 \qquad ...(7)$$

can serve as a model for reactions involving more complex hydrocarbons, including diamondoid moieties. The energy barrier for this process is substantial, ~100 maJ, making it representative of relatively difficult abstraction reactions.

Abstraction in the extended LEPS model

The piezochemistry of abstraction reactions can be modeled using the extended LEPS potential. This three-body potential automatically fits the bond energies, lengths, and vibrational frequencies of each of the three possible pairwise associations of atoms. In a symmetrical process, the two independent Sato parameters can be used to fit the barrier height and the CC separation, *r*(CC), at the linear, three-body transition state [calculated to be 0.2669 nm for reaction using high-order *ab initio* methods]. Fitting a LEPS function to these values using parameters based on standard bond lengths and energies for methane and ethane and bond-stretching stiffnesses from MM2, yields Sato parameters of 0.132 for the two CH interactions and –0.061 for the CC interaction. (The pure LEPS potential predicts deflection of the hydrogen atom from the axis at small CC separations, but a modest angle-bending stiffness suffices to stabilize a linear geometry.)

With this function in hand, it is straightforward to evaluate the energy barrier $\Delta V^{\ddagger}$ as a function of the compressive load F_{compr}. In this model, $\Delta V^{\ddagger} = 0$ at $F_{compr} \approx 3.6$ nN, and *r*(CC) ≈ 0.242 nm; this load is well within the achievable range. At ordinary temperatures, however, there is little practical reason to drive $\Delta V^{\ddagger}$ to zero (indeed, with tunneling, there is little reason to do so even at cryogenic temperatures). The barrier is reduced to $2kT_{300}$ at a load of 2 nN, with *r*(CC) ≈ 0.251 nm at the transition state. At a load of 1 nN, $\Delta V^{\ddagger} \approx 29$ maJ, and *r*(CC) ≈0.258 nm. Assuming a freuency factor of 10^{13} s^{-1}, this barrier would be consistent with an omitted-reaction probability $\leq 10^{-15}$ in a mechanochemical system with a transformation time of $\sim 4 \times 10^{-9}$ s (if, that is, the reaction were also sufficiently exoergic). These conditions are consistent with achieving high reaction reliability via the conditional-repetition mechanism.

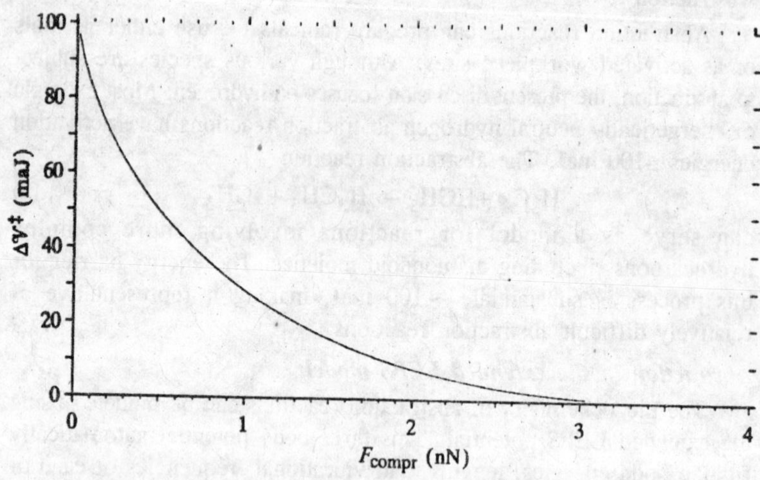

Fig. 9.5. Barrier height for abstraction of hydrogen from methane by methyl, plotted as a function of the compressive load applied to the carbon atoms.

Exoergic abstraction reactions

A reliable single-step reaction must meet stringent exoergicity requirements at the time of kinetic decoupling. The energy differences are consistent with $P_{err} < 10^{-15}$ without using cycles of repetition and measurement, while $P_{err} \approx 10^{-13}$. The energies in these reactions are computed from differences in C—H bond strengths.

Hydrogen abstraction tools

The large C—H bond strength of alkynes (~915 maJ) enables alkynyl radicals to abstract hydrogen atoms from most exposed nonalkyne sites with a single-step $P_{err} < 10^{-15}$. Alkyne moieties can be buttressed by nonbonded contacts to increase their stiffness, greatly reducing the probability of reacting with sites as much as a bond length from the target. In a typical situation, a supporting structure need only have good stiffness in resisting displacements in one direction, and can permit larger excursions in directions that lack nearby reactive sites.

Moieties with C—H bond energies between those of alkynes and aryls can abstract hydrogen atoms from most Sp^3 carbon sites with good reliability. Strained alkenyl radicals should bind hydrogen with the necessary energy (≥790 maJ, vs. ~737 maJ for ethenyl). These structures can exhibit good stiffness without buttressing from nonbonded contacts. Their steric properties are desirable, and their supporting structures strongly suppress bond-cleavage instabilities that would otherwise be promoted by the radical site.

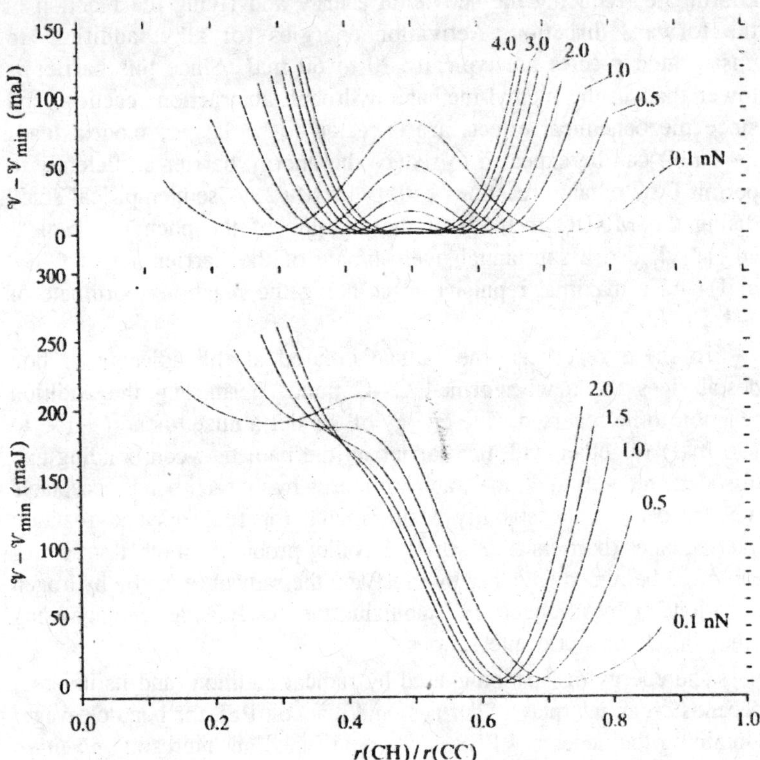

Fig. 9.6. Potential energy surfaces for abstraction reactions under various compressive loads, plotted as a function of a reaction coordinate, the ratio of a CH distance to the CC distance.

Hydrogen donation tools

Structures capable of resonant stabilization of a radical permit easy abstraction of a hydrogen; these include cyclopentadiene, and various other unsaturated systems. Some atoms bind hydrogen weakly, including tin and many other metals. Moieties like these, with C—H bond energies less than ~530 maJ, can donate hydrogen atoms reliably to carbon radicals where the product is an ordinary sp^3 structure. Accordingly assumes that the abstraction and donation of hydrogen atoms can be performed at will on diamondoid structures.

Alkene and Alkyne Radical Additions

The addition of radicals to alkenes and alkynes often is exoergic by ~160 to 190 maJ; this is sufficient to ensure reaction reliability. Because radical addition converts a (long) nonbonded contact into a (short) bond, compressive loads can couple directly to the reaction coordinate, reducing the activation energy and riving the reaction in the forward direction. Activation energies for alkyl additions to unsaturated species are typically 30 to 60 maJ. Since this barrier is lower than in the methyl-methane hydrogen abstraction reaction, and since piezochemical effects are expected to be larger, modest loads (~1 nN) can be expected to reduce the energy barrier sufficiently to permit fast, reliable reactions with $t_{trans} = 10^{-7}$ s. A semiempirical study (using the MNDO method) of the addition of the phenyl radical to ethene yielded a substantial overestimate of the barrier height (~98 maJ) and a maximal repulsive force along the reaction coordinate of ~1.5 nN.

In these reactions, the radical created at the adjacent carbon destabilizes the newly formed C—C bond, permitting the addition reaction to be reversed. The energy of the destabilized bond (~160 to 190 maJ), together with the barrier on the path between bonding and dissociation (~30 to 60 maJ, above), makes many such bonds acceptable with respect to the stability requirements for reagents and reaction intermediates (barriers ≥ 230 maJ). Stability problems in other structures can often be remedied either by satisfying the radical (e.g., by hydrogen donation) or by mechanically stabilizing the newly added moiety using other bonds or steric interactions.

The energy dissipation caused by radical addition (and its inverse) depends on an interplay of stiffness and reaction PES for bond cleavage. Obtaining the necessary PES data would make an interesting *ab initio* study.

Stiffness and misreaction rates in radical addition

Radical addition reactions provide a convenient example of a process having two similar, closely spaced transition states. Structure illustrates a transition state in the attack of an aryl radical on an alkene, with asterisks marking equivalent, alternative locations for the position of the attacking radical. Although transition states can be arbitrarily close together, the separation of the pair of locations on a single side of the alkene (moderately more than a bond length) is toward the low end of the range of anticipated typical cases, presenting a relatively challenging situation for achieving reliable positional control.

Quantum calculations can be used to estimate the transition-state geometry for the addition of a phenyl radical to ethene, yielding lengths of 0.135 nm for the breaking double bond and 0.225 nm for the bond undergoing formation, with an angle of 95.4^0 between them. This yields a distance between alternative positions for the attacking radical of

0.177 nm. (In the full configuration-space picture, the distance between transition states is increased by contributions from other displacements, including differences in the angle of the attacking radical.)

At this distance, an elastic energy difference of > 180 *maJ* can be achieved with a stiffness of ~12 N/m; a stiffness in this range can effectively suppress misreactions (rates $\leq 10^{-15}$). Because the most probable misreactions result from a displacement in a particular direction, however, an advantage can be gained by biasing the equilibrium position of the attacking moiety in the opposite direction during the approach to the reaction site. For example, with a stiffness of 6 N/m, a 0.1 nm bias of this sort results in an elastic energy difference of ~200 maJ between the two transition state geometries, while raising the energy of the favoured transition by only 25 mal Accordingly, stiffnesses in the 5 to 10 N/m range should prove adequate to suppress misreactions with high reliability in a wide range of synthetic operations.

Pi-bond Torsion

Rotation of one of the methylene groups of ethene by 90⁰ breaks the pi bond, yielding a diradical. Analogous structures and transformations can be used to modulate the strength of adjacent sigma bonds. The transition (a) → (b) represents an R—C bond cleavage yielding a twisted alkene; (c) → (d) represents an R—C cleavage yielding a planar alkene. The differences in alkene energy around the

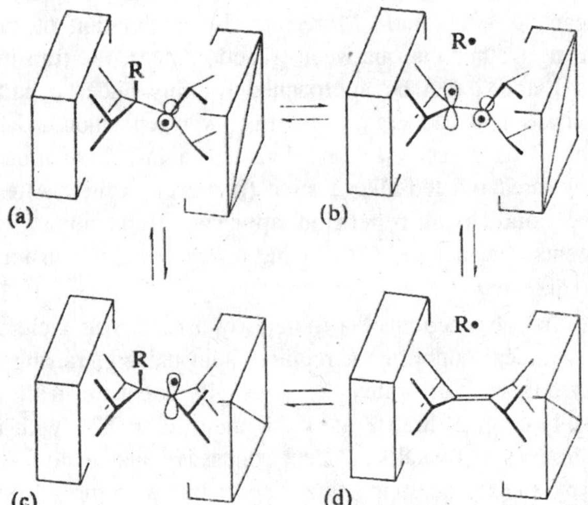

Fig. 9.7. Modulation of bond strength by pi-bond torsion.

illustrated cycle can be used to estimate the difference in R-group bond energies between (a) and (b). The (a) → (c) transition involves torsion of a single bond that links centers of roughly twofold and threefold symmetry; the resulting energy difference is usually small in the absence of substantial steric interference between the end groups. The difference in R-group bond energies between (a) and (c) thus approximates the energy difference between twisted and planar alkenes (b) and (d). The analogous energy difference is 453 maJ for ethene. Alternatively, one can assume that the R-group bond energy in a structure like (a) is unaffected by the presence of an adjacent *twisted* radical (i.e., would be unchanged if the radical site were hydrogenated), then use the bond weakening caused by an *unconstrained* radical to estimate the energy difference; from thermochemical data, this difference is ~410 to 440 maJ for hydrocarbon R-groups. Thus, the bond energy of a typical R-group in (a), ~550 to 650 maJ, drops dramatically when the structure is twisted to the configuration of (c). Indeed, elimination of a more weakly bonded group at this site can become exoergic.

A typical reaction cycle for a mechanism of this kind could involve (1) abstraction of a relatively tightly bound moiety by a radical site like that in (b), yielding a structure like (a); (2) torsion to a state like (c) weakening the new bond; and (3) abstraction of the moiety by another radical, delivering it to a more weakly bound (i.e., high-energy) site and leaving a structure like (d). If steps 1 and 2 must be exoergic by 145 maJ, then the net increase in the energy of the transferred moiety can be ≥120 maJ. Moreover, by modulation of the torsion angle during (rather than between) reaction steps, the transition states for steps 1 and 2 can be approached forcibly under conditions that make the transitions endoergic or isoergic, with separation under exoergic conditions. This meets the conditions for a low-dissipation process, and the isolated twisted-alkene state (b) never occurs. Alternatively, using the conditional repetition approach to remove exoergicity requirements, the net increase in the energy of the transferred atom can be >400 maJ.

Note that abstraction of a moiety by a radical to yield an alkene resembles radical coupling: it requires spin pairing, raising questions of intersystem crossing rates. k_{isc} is usually adequate in the presence of a suitably coupled high-Z atom. Bismuth ($Z = 83$), with its ability to form three (albeit weak) covalent bonds, is a candidate for inclusion at a nearby site in the supporting diamondoid structure.

Mechanochemical processes involving pi-bond torsion appear to have broad applications. For example, reactions of dienes can be accelerated by torsions that weaken the initial pair of double bonds. In the reverse direction, the sigma bonds resulting from the reaction can (for reactions yielding noncyclic products) later be cleaved with the aid of radicals generated by torsion of the new double bond. Some other systems having a single pi bond can likewise be manipulated by mechanical torsion; these include some metal-ligand and boron-nitrogen double bonds. In the latter instance, torsion results in a lone pair and an empty orbital, causing a shift in electron density.

Radical Displacements

Various mechanochemical processes are analogous to pi-bond torsion in that they modulate the strength of a sigma bond by altering the availability of a radical able to form a competing pi bond. For example, the addition of a radical to an alkyne can facilitate an abstraction reaction.

A process of this sort could employ a weakly bonding moiety (or one with modulable bonding) as the attacking radical R, generating a strong alkynyl radical by tensile bond cleavage. A mechanochemical cycle based on these steps can first use an alkynyl radical to abstract a tightly bound atom from one location, then deliver the atom to a moderately bound site R', and finally regenerate the original alkynyl radical by cleaving a weak bond to R. Given the strong weakening of the bond to hydrogen in (b), each of these steps should proceed rapidly and reliably under moderate mechanical loads. Analogous displacement operations can be used to regenerate alkene- and aryl-derived sigma radicals.

Carbene Additions and Insertions

The standard carbene addition and insertion reactions have low barriers. The changes in geometry resulting from these reactions show that mechanical loads can be directly coupled to the reaction coordinate, producing strong piezochemical effects; the increase in bond number points to the same conclusion. The insertion of carbenes into C—C bonds, although analogous to metal insertion reactions, has not been observed in solution-phase chemistry; the ubiquitous presence of alternative reaction pathways with less steric hindrance, lower energy barriers, or greater exoergicity is presumably responsible, since this reaction is exoergic and permitted by orbital symmetry. It presents a reaction step which assumes that positional control and mechanical

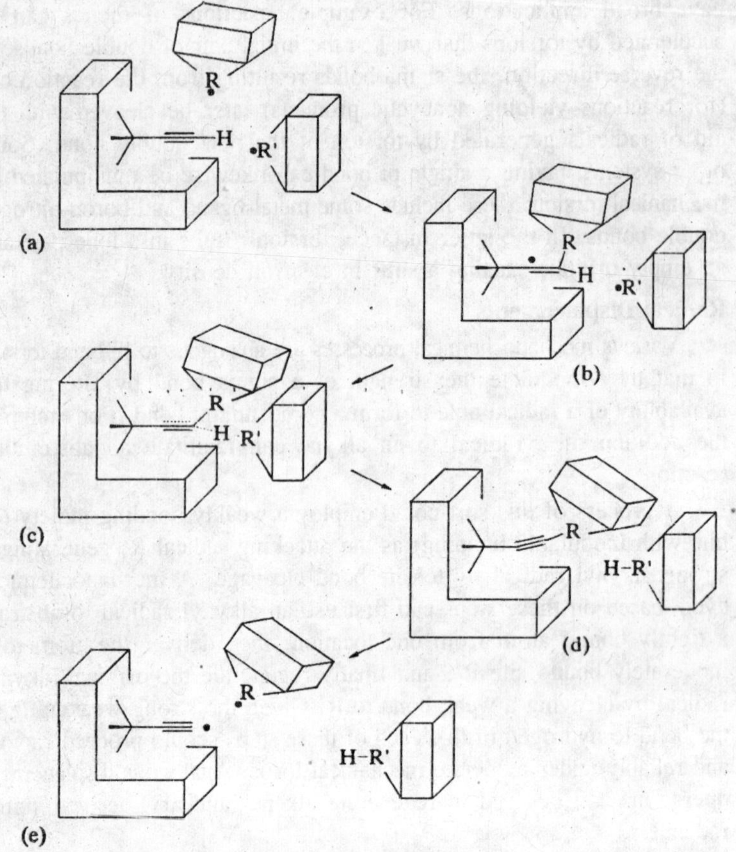

Fig. 9.8. Removal of a tightly bound alkynyl hydrogen atom, facilitated by the addition of a radical

forces can effect carbene insertion into a strained, sterically exposed C—C bond.

Since bond-forming carbene reactions are highly exoergic, single-step reactions can be highly reliable (assuming, as always, that conditions and mechanical constraints are chosen to exclude access to transition states leading to unwanted products). Fast, reliable reactions involving triplet carbenes can require fast intersystem crossing. The potential for low-dissipation carbene reactions is at present unclear.

Alkene and Alkyne Cycloadditions

°Cycloaddition reactions can find broad use in mechanosynthesis, as they have in solution-phase synthesis. The Diels-Alder reactions form two bonds and a ring simultaneously, and, as has been mentioned, they have large negative volumes of activation at moderate pressures (~-0.05 to -0.07 nm^3). These reactions are more sensitive to piezochemical effects than are abstraction reactions. Energy barriers vary widely, from <60 to >130 maJ. Reaction proceeds at low rates at ordinary temperatures and pressures; piezochemical rate accelerations under modest loads should make them consistent with $t_{trans} = 10^{-7}$ s. Under these conditions, alkenes can be replaced with less-reactive alkynes, yielding less saturated products. Exoergicity is sufficient to yield stable intermediate products in a reliable process.

Cycloaddition reactions are subject to strong orbital-symmetry effects. The [2 + 2] cycloaddition reaction is termed ***thermally forbidden*** (according to the Woodward-Hoffmann rules), because the highest-energy occupied orbital on one molecule fails to mesh properly with the lowest-energy unoccupied orbital on the other (the unoccupied orbital presents two lobes of opposite sign, divided by a node; the occupied orbital has no such node, and creating the required node is akin to breaking a bond). But, as with well-observed "forbidden" spectral lines, the ban is not complete. Piezochemical techniques (including pi-bond torsion and direct compression) can force the formation of one of the new sigma bonds, even at the expense of breaking both pi bonds, thereby creating a pair of radicals that can then combine to form the second sigma bond. This reaction can also proceed through a polar, asymmetric intermediate, which is likewise subject to strong piezochemical effects. For the dimerization of ethene, the activation energy is ~305 maJ.

Transition-metal Reactions

The present section describes some reactions and mechanochemical issues in more detail.

Reactions involving transition metals, carbon, and hydrogen

Transition metals participate in a wide variety of reactions, including many that make or break carbon-carbon and carbon-hydrogen bonds. All illustrated states and transformations in each sequence have been observed, though not necessarily in a single complex. The complexes have single bonds between metal and carbon, but double-bonded species (metal carbene complexes and alkylidenes) and triple-bonded species (metal alkylidynes) are also known.

Species such as metal alkylidene carbenes and metal alkylidyne radicals can presumably exist (given stable ligands and a suitable eutactic environment) and may be of considerable use in synthesis.

(Note that high-Z transition metals can accelerate intersystem crossing in reactions in which they participate.)

Among the reactions of metal carbene complexes are the following. Again, all illustrated states and transformations in each sequence have been observed, though not necessarily in a single complex. Transition metals can also serve as radical leaving groups in S_H2 reactions at sp^3 carbon atoms, transferring alkyl groups to radicals.

Ligands suitable for mechanochemistry

The reactivity of a transition metal atom is strongly affected by the electronic and steric properties of its ligands. These can modify the charge on the metal, the electron densities and energies of various orbitals, and the room available for a new ligand. A ligand can be displaced by a new ligand, or can react with it and dissociate to form a product.

In a molecular manufacturing context, reagent stability is commonly adeqate if all components are bound with energies ≥230 maJ. Typical M—C bond strengths are ~210 to 450 maJ (vs. ~550 maJ for typical C—C bonds); the stronger bonds have adequate stability by themselves, and weaker bonds are acceptable if incorporated into a stabilizing cyclic structure. Stabilization by cyclic structures will often be necessary to prevent unwanted rearrangements in the ligand shell. Since M—H bonds are estimated to have strengths ~100 to 200 maJ greater than M—C bonds, their strengths should be adequate in the absence of special destabilizing circumstances. Electronegative ligands such as F and Cl should also be relatively stable, particularly in structures that lack adjacent sites resembling anion solvation shells in polar solvents.

The stability and manipulability of ligand structures in a mechanochemical context are greatest when they are bound to strong supporting structures. This is easily arranged for a wide variety of ligands having metal-bonded carbon, nitrogen. oxygen, phosphorus, or sulfur atoms. Of these, phosphorus (in the form of tertiary phosphines, PR_3) has been of particular importance in transition-metal chemistry. Carbon monoxide, another common ligand, lacks an attachment point for such a handle and may accordingly be of limited use. Isonitriles, however, are formally isoelectronic to CO at the coordinating carbon atom; they exhibit broadly similar chemistry and can be attached to rigid, extended R groups. (Many ligands not mentioned here also have useful properties.)

Ligand supporting structures can maintain substantial strength and stiffness while occupying a reasonably small solid angle. This enables several ligands to be subjected to simultaneous, independent mechanochemical manipulation. The MM2/CSC stiffness of the following diamondoid support structure is ~115 N/m for extension and ~20 N/m for bending, both measured for displacements of the carbon atom at the tip relative to the bounding hydrogen atoms at the base.

Alternatively, several ligand moieties can be integral parts of a diamondoid structure bound metal atom with two of six octahedral coordination sites. The metal-nitrogen bond lengths shown in this structure are appropriate for octahedrally coordinated cobalt.

Single ligands mounted on independently manipulable tips represent one extreme of mobility; multiple ligand moieties in a single rigid structure represent another. An intermediate class incorporates several ligand moieties into a single structure, subjecting them to substantial relative motion by elastic deformation of that structure. This can ensure large interligand stiffnesses, facilitating low-dissipation processes.

Mechanically driven processes

Low-stiffness, low-strength bonds are more readily subject to mechanochemical manipulation: both the required force and the required stiffness of the surrounding structure are smaller. The stretching frequencies of M—H bonds imply stiffnesses between 130 and 225 N/m, ~0.25 to 0.50 times the stiffness of a typical C—H or C—C bond; the stiffness of M—C bonds (which are longer and of lower energy) should likewise be low. Given that bonds as stiff as C—C can be cleaved in a low-dissipation process, a wide range of mechanochemical processes involving transition metals can presumably be carried out in a positive-stiffness, zero-barrier manner (or, with similar effect, in a manner encountering only small regions of negative stiffness, and hence only low barriers).

In solution-phase chemistry, catalysts capable of inserting metals into alkane C—H bonds have been unstable, either attacking their ligands or the solvent, and insertion into C—C bonds has required a strained reagent. With an expanded choice of ligands and the elimination of accessible solvent molecules, the first problem should be avoidable. Further, when loads of bond-breaking magnitude can be applied between a transition-metal atom and a potential reagent, intrinsic strain is presumably no longer required in the reagent. Configurations seem well suited for insertion into a sterically exposed bond. Since metal insertion in the above instances has the effect of replacing a strong, stiff bond with two weaker, more compliant bonds, it can facilitate further mechanochemical operations.

Transition-metal complexes with large coordination numbers lend themselves to bond modulation based on manipulation of steric crowding. In octahedral complexes, for example, the metal atom can be anchored by (say) three ligands while two of the remaining ligands are rotated or displaced to modulate overlap repulsion on the sixth. By analogy with processes observed in solution, the introduction of new ligands can be used to expel existing ligands by a combination of steric and electronic effects. Mechanochemical processes can forcibly introduce ligands almost regardless of their chemical affinity for the metal, driving the expulsion of other, relatively tightly bound ligands. Conversely, such processes can remove ligands (having suitable "handles"), even when they are themselves tightly bound. Again, the presence of multiple other ligands to anchor the metal atom facilitates such manipulations.

More subtly (and conventionally), the binding of a ligand in a square or octahedral complex can be strongly affected by the electronic properties of the ligand on the opposite side (the *trans*-effect); changes in this *trans* ligand can alter reaction rates by a factor of $\sim 10^4$, suggesting changes in transition-state energy of ~ 40 maJ. Accordingly, mechanical substitution or other alteration of ligands (e.g., double-bond torsions) should be effective in modulating bonding at *trans* sites in mechanochemical reaction cycles.

In summary, although transition metals are of only moderate interest as parts of nanomechanical products, they are of considerable interest as components of tools in mechanochemical systems for building those products. Their comparatively soft interactions, relaxed electronic constraints, and numerous manipulable degrees of freedom suit them for the preparation and recycling of reagent moieties, and (where steric constraints can be met) for direct use as reagents in product synthesis. In light of the broad capabilities of other reagents under mechanochemical conditions, the use of transition-metal reagents is unlikely to expand the range of structures that can be built; it is, however, likely to expand greatly the range of structures that can be built with low dissipation, in a nearly thermodynamically reversible fashion.

Mechanosynthesis of Diamondoid Structures

Fundamental physical considerations (strength, stiffness, feature size) favour the widespread use of diamondoid structures in nanomechanical systems. In chemical terms, diamondoid structures comprise a wide range of polycyclic organic molecules consisting of fused, conformationally rigid cages. This section considers the synthesis of such structures by mechanochemical means, based on reagents and processes of sorts described in the preceding section, and using diamond itself as an example of a target for synthesis.

Why Examine the Synthesis of Diamond?

Diamond is an important product in its own right, but here serves chiefly as a test case in exploring the feasibility of more general synthesis capabilities. It is impractical at present to examine in detail the synthesis of numerous large-scale structures. Accordingly, it is important to choose a few appropriately challenging models.

Diamond has several advantages in this regard, as can be seen by a series of comparisons. Synthetic challenges often center around the framework of a molecule, and diamond is pure framework. In general, higher °valence and participation in more rings makes an atom more difficult to bond correctly. At one extreme is hydrogen placement on a surface; at the other is the formation of multiple rings through tetravalent atoms. Solid silicon and germanium present the same topological challenges as diamond, but atoms lower in the periodic table are more readily subject to mechanochemical manipulation owing to their larger sizes and lower bond strengths and stiffnesses. Thus, a structure built entirely of rings of Sp^3 carbon atoms appears to maximize the basic challenges of bond formation, and diamond is such a structure. Further, diamond has the highest atom and covalent-bond density of any well-characterized material at ordinary pressures, maximizing problems of steric congestion. Although diamond is a relatively low-energy structure, lacking significant strain or unusual bonds, existing achievements suggest that the latter features need not be barriers to synthesis, even in solution-phase processes.

Finally, diamond is a simple and regular example of a diamondoid structure. Accordingly, the description of a small synthetic cycle can suffice to describe the synthesis of an indefinitely large object; this avoids the dilemma of choosing between (1) syntheses too complex to describe in the available space, and (2) syntheses that might in some way be limited to small structures.

Why Examine Multiple Synthesis Strategies?

The identification of several distinct ways to synthesize a particular structure suggests that ways can be found to synthesize different but similar structures. Identifying multiple syntheses for diamond provides this sort of evidence regarding the synthesis of other diamondoids. Further, diamond itself has several sterically and electronically distinct surfaces on which construction can proceed, and sites on these surfaces can serve as models for the diverse local structures arising in the synthesis of less regular diamondoids. Accordingly, the following sections survey several quite different techniques, not to buttress the case for diamond synthesis (a process already known in the laboratory), but to explore the capabilities of mechanosynthesis for building broadly diamondlike structures by using diamond as an example.

Diamond Surfaces

Corners and other exposed sites pose fewer steric problems than sites at steps in the middle of planar surfaces. This section considers a set of reaction cycles at such sites on low-index diamond surfaces; the present section introduces the surfaces themselves. (In the diagrams here and in the following section, structures are truncated without indicating bonds to missing atoms.)

The (111) surface

When prepared by standard grinding procedures, a losely packed diamond (111) surface is hydrogen terminated. When heated sufficiently to remove most of the hydrogen (1200 to 1300 K), this surface reconstructs into a structure with (2 × 1) symmetry; the Pandey pi-bonded chain model for this surface. One calculation yielded an exoergicity of ~50 maJ per surface atom for this reconstruction, driven by the conversion of radical electrons to bonding electrons.

It is not presently known whether a bare diamond (111) surface, it spontaneously transforms to the (2 × 1) structure at room temperature. Calculations for the similar reconstruction of a silicon (111) surface suggest an energy barrier of ≤5 maJ per surface atom. This energy barrier, however, cannot be equated with an energy barrier for the nucleation of a (2 × 1) domain on an unreconstructed (1 × 1) surface, since nucleation requires the simultaneous motion of a number of atoms and stores deformation energy in the domain boundary. Whatever the true situation, since hydrogenation is known to suppress the (2 × 1) reconstruction, and since donating and abstracting hydrogen atoms from (111) surface sites is straightforward, an unreconstructed (111) surface can be maintained during construction.

The (110) surface

Termination of the bulk structure leads to a surface. The bonded chains along the surface resemble those formed in the (2 × 1) reconstruction of a (111) surface, but with the pi bonds subject to greater torsion and pyramidalization. Although this geometry generates no radicals, the pi bonds are quite weak, and hence the p-orbital electrons can exhibit radical-like reactivity.

The (100) surface

Simple truncation of the bulk structure would yield a surface covered with carbene sites. Displacement of surface atoms to form pi bonded pairs yields the stable reconstruction. The resulting surface alkene moieties are strongly pyramidalized and under substantial tension, increasing their reactivity.

Some chemical observations

The diversity of surface structures possible on a single bulk structure, diamond, shows how a diamondoid structure (like most complex molecules) can be built up through intermediates having widely varying chemical properties. Further, the differing strained-alkene moieties found on (110) and (100) surfaces show that stable diamondoid intermediates can have highly reactive surfaces, facilitating synthetic operations. Finally, the (possible) instability of a bare, radical-dense, unreconstructed (111) surface illustrates how requirements for temporary, stabilizing additions (such as bond-terminating hydrogens) can arise.

Stepwise Synthesis Processes

Synthesis strategies that has described already exploit the regular structure of diamond by laying down reactive molecular strands. Synthesis based on the mechanical placement of small molecular fragments, in contrast, suggests how specific *irregular* structures might be synthesized, and thus provides a point of departure for considering the synthesis of a broad class of diamondoids.

Existing models of diamond synthesis

Models of synthesis via small molecular fragments have been developed to explain the low-pressure synthesis of diamond under non equilibrium conditions in a high-temperature hydrocarbon gas, a process of increasing technological importance. Two models have been advanced and subjected to studies using semiempirical quantum mechanics. Both propose mechanisms for the growth of diamond on (111) surfaces, one ased on a cationic process involving methyl groups and one based on the addition of ethyne. Of these, the ethyne process appears more plausible and more directly relevant to feasible mechanosynthetic processes (regardless of its frequency during diamond growth from the gas phase).

The addition of two ethyne molecules to a hydrocarbon molecule which, with suitable positional constraints, was used to model a step on a diamond (111) surface. The overall set of calculations used MNDO and consumed ~35 hours of CPU time on a Cray XMP/48. The reaction mechanism originally proposed for the transformations 2 → 4 and 4 → 6 involved multiple steps, later shown to be concerted. The initiation step involves abstraction of a hydrogen atom by atomic hydrogen (1 → 2), and has a significant energy barrier. Steps were calculated to proceed without energy barriers, suggesting that any energy barriers that actually occur are unlikely to be large.

In an analogous mechanosynthetic process, ethyne could be replaced by an alkyne moiety bonded to a structure serving as a handle. Step 5 liberates a free hydrogen atom, which is unacceptable in a eutactic environment, but a related approach avoids this loss of control.

Mechanosynthesis on (111)

Step 1 of a kink site in a step on a hydrogenated diamond (111) surface; one bond results from a reconstruction, and two hydrogen atoms have been removed to prepare radical sites. In step 2 of the proposed synthetic cycle, an alkyne reagent moiety is applied to one of the radical sites, resulting in a transition structure analogous to step 3. The chief differences are that insertion occurs into a strained C—C bond rather than an unstrained C—H bond, and that large mechanical forces can (optionally) be applied to drive the insertion process. In step 3, tensile bond cleavage occurs (use of Si or another non-first-row atom reduces the mechanical strength of this bond), and a bond forms to the remaining prepared radical. Deposition of a hydrogen at the newly generated radical site and abstraction of two hydrogens further along the surface step then completes the cycle (not shown), generating a kink site like that in step 1, but with the diamond lattice extended by two atoms. Save for accommodating boundary conditions at the edge of a finite (111) surface, this sequence of operations (like those in the following sections) suffices to build an indefinitely large volume of diamond lattice.

Mechanosynthesis on (110)

The extension of a pi bonded chain along a groove in a diamond (110) surface, using an alkylidenecarbene reagent moiety. Step 1 illustrates the starting state; step 2 illustrates a transitional state in the insertion of the carbene into a strained C—C bond. This process takes advantage of mechanical constraints to prevent the addition of the carbene into the adjacent double bond to form a cyclopropane moiety. Instead, electron density is accepted from the terminal atom of the pi bonded chain into the empty *p*-orbital of the carbene carbon, developing one of the desired bonds, while electron density is donated from the sigma orbital of the carbene carbon to form the other desired bond. Substantial forces (≥4 nN) can be applied to drive this process, limited chiefly by mechanical instabilities.

Step 4 illustrates the use of torsion to break a pi bond, thereby facilitating tensile bond cleavage, step 5. The final state, step 6, resembles that in step 1 save for the extension of the lattice by one atom. A further cycle (restoring the state of step 1 exactly, save for the addition of two atoms) would be almost identical, except that the equivalent of step 5 involves attack by a newly forming radical on a weak pi bond, rather than its combination with an existing radical.

Mechanosynthesis on (100)

A synthetic cycle on diamond (100) is illustrated, in which reactions occur on a series of relatively independent rows of pairs of dimers. In step 1, a strained cycloalkyne reagent moiety is applied. (The division of the supporting structure into blocks is intended to suggest opportunities for modulating bond strength by control of torsional deformations; alternative structures could serve the same role, providing weak bonds for later cleavage.) The reaction in step 1 (promoted by the nearly diradical character of the strained alkyne and by applied mechanical force) is formally a cycloaddition process, provided that the resulting pair of radicals is regarded as forming a highly elongated pi bond. Step 2 is then formally a (thermally forbidden) cycloaddition process, but the small energy difference between the bonding and antibonding orbitals in the "pi bond" provides a low-lying orbital of the correct symmetry for bonding, hence the forbiddenness should be weak. Moderate mechanical loads should suffice to overcome the associated energy barrier. Step 4 is an endoergic retro-Diels-Alder reaction yielding a high-energy, highly pyramidalized alkene moiety (which is, however, less pyramidalized than cubene). The energy for this process is supplied by mechanical work.

Step 5 represents the state of the row after bridging dimers have been deposited at all sites. The resulting three-membered rings are analogous to epoxide structures in a model of an oxidized diamond (100) (2 × 1) half-monolayer surface.

Steps 6-8 represent a cycle in which dimers are sequentially inserted along the row: In step 6, a dimer has already been added to the right, cleaving the strained rings and generating two radical sites adjacent to a cleft in the surface. Step 7 illustrates the bonds that undergo formation and cleavage as a result of the mechanical insertion of a strained alkyne into the cleft. The nature of the transition state will depend on the spin state of the radical pair and orbital symmetry considerations. The large exoergicity for this process can be seen from a comparison of the bonds lost (two pi bonds in the strained alkene and two strained sigma bonds in the three-membered rings) with the bonds formed (four almost strain-free sigma bonds). The geometry of the cleft permits the application of large loads without mechanical instability, hence large energy barriers to bond formation (in the unloaded state) are acceptable.

Strand Deposition Processes

Stepwise synthetic processes need not be applied in regular cycles to build up a regular structure, but could instead be orchestrated to build up diamondoid structures tailored for specific purposes. Where diamond itself is the target, synthesis can take direct advantage of structural regularities.

Cumulene strands

Cumulene strands are high-energy, pure-carbon structures that represent promising precursors in the mechanosynthesis of diamond. It represents a pair of similar reaction processes on a dehydrogenated step on a diamond (111) surface and on a groove in a (110) surface: in each. a pi-bonded chain is extended by the formation of additional bonds to a cumulene strand in an exoergic, largely self-aligning process. Substantial forces can be applied through nonbonded interactions (which can also be used to constrain strand motions). The reaction on a hydrogenated (111) surface may require a series of stepwise reactions both to dehydrogenate atoms in the plane to be covered and to hydrogenate new atoms in the plane being constructed; the (110) reaction has no such requirement.

Hexagonal diamond from hexagonal strands

The unsaturated, hexagonal, columnar structure can be regarded as a tightly rolled tube of graphite; it can made from a °saturated structure by abstraction of all hydrogens. Like a cumulene strand, this is a pure-carbon structure; owing to pyramidalization and torsion of pi systems, it is also relatively high in energy.

A semiempirical quantum chemistry study using AMI on the model structure yielded bond lengths of ~0.1384 nm for the three central, axially aligned bonds, and ~0.1510 nm for the twelve adjacent bonds. These values are close to standard values for pure double and pure single bonds (0.1337 and 0.1541 nm), hence structure representing the hexagonal column as a network of strained alkenes, provides a good description.

Figure represents a (100) surface of hexagonal diamond, bounded by similar strained alkenes. Figure illustrates a reaction in which a hexagonal column bonds to a groove adjacent to a step on that surface; this can be regarded as proceeding by the attack of a strand radical on a surface alkene, generating a surface radical, which then attacks the next strand alkene, and so forth. Thus, each row of alkenes undergoes a process directly analogous to the free-radical chain polymerization that yields polyethylene, save for the greater reactivity of the participating alkenes and the presence of mechanical forces tending to force each radical addition.

Cluster-based Strategies

Syntheses based on cumulene and hexagonal-column strands suggest the feasibility of synthesizing diamondoid solids using reactive molecular fragments of intermediate size (e.g., 10 to 30 atoms). Fragments of this size can be strongly convex, relaxing steric constraints in their synthesis. Containing tens of atoms, they can embody significant structural complexity and deliver that preformed complexity to a workpiece. By incorporating unsaturated structures, radicals, zarbenes, and the like, they can form dense arrays of bonds to a complementary surface. Fullerenes are examples of unsaturated carbon clusters, and C_{60} has been crushed to diamond at room temperature by anisotropic compressive loads of ~20 GPa, suggesting the feasibility of mechanosynthesis of diamond and diamondoid structures from fullerene precursors.

This approach is a form of the familiar chemical strategy of convergent synthesis. Further, the contemplated size range of these fragments is familiar in organic synthesis today; their relatively high reactivity could be achieved (starting with more conventional structures) by a series of mechanochemically guided abstraction reactions in the protection of a eutactic environment.

Toward less Diamondlike Diamondoids

Members of the broad class of diamondoids can differ from diamond both in patterns of bonding and in elemental composition. The differing bonding patterns created during intermediate stages of the diamond syntheses proposed in this section, together with general experience in chemistry, suggests that nondiamond structures are readily accessible. Deviations from the diamond pattern usually reduce the overall number density of atoms, thereby reducing steric congestion and (all else being equal) facilitating synthesis.

Regarding differences in elemental composition, it is significant that the classes of reagent species exploited in this section (unsaturated hydrocarbons, carbon radicals, and carbenes) have analogues among other chemical elements. For example, most other elements of structural interest (N, 0, Si, P, S), all can form double (and sometimes triple) bonds [e.g., C=N, C≡N, C=O, C=S; and the more exotic examples Si=Si and C=Si, and C=P and C≡P]. All these elements can host radical sites. Silicon can form a carbenelike divalent species, silene. Fluorine, chlorine, and bromine can, like hydrogen, participate in abstraction reactions. The focus on hydrocarbon structures in this chapter has been driven more by limits on time and page space than by limits on chemistry.

The diversity of feasible syntheses for a challenging test-case, diamond, suggests that most reasonably stable diamondoid structures will prove susceptible to mechanosynthesis.

Mechanosynthesis of Nondiamondoid Structures

As we move away from diamondoids, while remaining within the class of reasonably stable structures, synthesis appears to grow easier for reasons of the sort discussed else where in this chapter. Within the class of covalent structures, reduced rigidity might hamper mechanosynthesis, but the feasibility of using bound intermediates (using either covalent bonds or nonbonded interactions) can largely compensate. Further, the synthesis of flexible covalent structures is in the mainstream of existing chemical achievement.

Noncovalent solids present different challenges. In general, however, the ability to transfer atoms one at a time or in small clusters, and to perform piezochemical manipulation on a growing surface, can provide broad control. Accordingly, it seems reasonable to assume that most reasonably stable structures—diamondoid or not—will prove susceptible to mechanosynthesis. Aside from considering the use of metallic conductors, the balance of this volume will neglect noncovalent structures.

Conclusions

The achievements of solution-phase synthesis show that a remarkably wide range of molecular structures can be built, despite the absence of the standard basis for construction, that is, the ability to move and position parts to guide assembly. A comparison reveals several limitations of mechanosynthesis relative to solution-phase synthesis, but displays a more-than-compensating set of strengths. Chief among these strengths are (1) the ability to achieve reactions at specific sites while avoiding them elsewhere by exploiting direct positional control of reagent moieties, (2) as a consequence of this, the ability to apply highly reactive moieties, such as radicals and carbenes, with great specificity, and (3) the ability to accelerate reactions by the application of localized forces of bondbreaking magnitude.

Misreaction probabilities are exponentially dependent on relative energies, and with reasonable barrier heights can be $<10^{-15}$; omitted-reaction probabilities are exponentially dependent on repetition and relative well depths, and can likewise be low. It appears that the required conditions for high reliability do not excessively constrain the set of feasible synthetic operations. As a consequence, precise structures having $> 10^{12}$ atoms can be made in good yield.

A review of reactive species used in solution-phase chemistry suggests that unsaturated hydrocarbons, carbon radicals, carbenes, and transition-metal complexes (among others) are attractive in a mechanosynthetic context. A consideration of the utility of these species when used in conjunction with positional control and mechanical force indicates the feasibility of a wide range of useful transformations, some of which can be performed with energy dissipation $<kT$.

Finally, the mechanosynthetic production of diamondoid structures has been examined, using several proposed syntheses of diamond as an example. The diversity of feasible reactions (and the availability of analogous reactions involving elements other than C and H) indicates the feasibility of constructing a wide range of diamondoid and other structures.

Some open problems

Mechanosynthesis presents numerous open theoretical problems. The analysis in this chapter indicates the feasibility of a wide range of processes, but it describes only a few in substantial structural and energetic detail. The broader application of *ab initio* molecular orbital methods to potential mechanosynthetic reactions is a high priority. Research is needed to identify and characterize reaction pathways in mechanosynthesis of diamond and diamondoid structures. Similar studies are needed to describe steps in reagent preparation. A major challenge in this work will be the identification and characterization of potential misreaction pathways to determine whether the proposed reaction can be made reliable enough for molecular manufacturing. Semiempirical molecular orbital methods can assist in the exploration of relevant potential energy surfaces.

A long-term objective in this area is to compile a large and well-characterized set of mechanosynthetic reactions suitable for the synthesis of diamondoid structures. This, in turn, can eventually provide the basis for software able to perform retrosynthetic analysis, accepting a component design as an input and producing a description of a corresponding mechanosynthetic process.

10

Nanoscale Components

Nanomechanical systems differ from systems of biological molecular machinery in their basic architecture: nanomechanical components are supported and constrained by stiff housings, while biological components often can move freely with respect to one another. It is natural to focus on moving parts, but a typical system has much of its mass in the form of a stiff housing. Gears, bearings, springs, screws, sliding rods, motors—each should be pictured as anchored to or embedded in an extended diamondoid structure tailored to support it in a functional position with respect to other components.

The analogy to macroscale engineering practice is clear: the components of macroscale machines and other systems are commonly mounted on an extended structure, whether this is termed a housing, chassis, casing, airframe, engine block, or framework. A housing structure forms a unit, but it need not be monolithic. Macroscale housings can be made from pieces held together by fasteners, adhesives, or welding; nanomechanical housings can likewise be made from pieces held together by fasteners or adhesive interfaces. Moving parts typically occupy or sweep out only a modest fraction of the volume enclosed by a housing; the rest, save for small clearances, can be occupied by a solid structure with a diamondlike modulus. Accordingly, housings can often be quite stiff. Note that the thermal positional uncertainty of one region of a homogeneous three-dimensional structure with respect to another approaches a distance-independent constant at large separations. Since the stiffness of a housing can often be much greater than that of the moving parts, a housing can often be treated as rigid on a large scale, making corrections (when necessary) for local compliance.

Materials and Models for Nanoscale Components

Macromechanical engineering treats components as being made from materials, neglecting the placement of specific atoms.

Classes of Materials

Conventional metallic solids

In common metals (e.g., iron, aluminum, copper), atoms have 8 to 12 nearest neighbours with which they share weakly directional bonds. A simple model of ^{o}valence electrons in a metal pictures them as forming a gas in which the ionic cores of the atoms are embedded; in this model, bonding is nondirectional.

Bonds of low directionality allow dislocations to move smoothly, enabling atomic layers to slide over one another with relatively little resistance. This nanoscale behaviour permits macroscale ductility, making objects insensitive to cracks and able to undergo large plastic deformations. Brittle materials, in contrast, resist dislocation motion; this makes them stronger, but unable to deform plastically and highly sensitive to microscopic flaws. This sensitivity occurs because cracks under tensile loads concentrate the stress at their tip: ductile materials deform to blunt the tip and reduce the stress; brittle materials break, extending the crack and raising the tip stress, leading to catastrophic failure.

Because conventional manufacturing makes objects with a high density of microscopic flaws, strong-but-brittle materials have proved difficult to use, and weaker-but-ductile metals have dominated macro scale engineering practice. Molecular manufacturing, however, can limit flaws to rare defects of atomic scale, enabling the wider use of brittle materials. (Other tactics, such as dividing materials into fibers, can also make brittle materials useful; see discussions in the extensive literature on fracture mechanics.)

The atomic mobility permitted by metallic bonding sharply limits the usefulness of metals in a nanomechanical context at ordinary temperatures. Although some refractory metals and intermetallic compounds may in fact be useful, this volume assumes no use of metals as mechanical components.

Ionic solids

An ideal ionic solid consists of ions held together almost exclusively by electrostatic forces. Because an ion of one charge is strongly attracted to its oppositely charged nearest neighbours, but repelled by more distant neighbours, smooth sliding is difficult, and brittle behaviour results. Since ionic solids have no great strength to compensate for their *brittleness*, they find no significant application in macromechanical engineering and are expected to find no significant application in nanomechanical engineering.

Some solids have mixed covalent and ionic bonding (e.g., alumina, silica) and high strength. These materials (ceramics and glasses) are sometimes regarded as ionic solids; for present purposes, they can be regarded as borderline covalent solids. They find use in macromechanical engineering (limited chiefly by their brittleness), and may find use in nanomechanical engineering (limited chiefly by the availability of covalent materials with better properties).

Molecular solids

Weak forces (e.g., van der Waals attraction) can bind inert gas atoms and neutral, covalently bound molecules to form a solid. Unless the molecules are large, molecular solids are weak and of no use as structural materials.

Polymeric solids

In a polymeric solid, weak forces provide effective cohesion between long molecules, and the covalent bonds within those molecules can dominate the *bulk* properties of the material. *Polymeric solids* with well-aligned molecules can be strong and stiff along the axis of alignment. Many polymeric solids are also tough and ductile; they find widespread use in macroscale engineering.

In nanomechanical engineering, even small-scale mobility of polymer chains must usually be suppressed. This can most directly be accomplished by provided covalent cross links between chains, but in the limiting case, a crosslinked polymer becomes a covalent solid.

Covalent solids

In a covalent solid, bonds form a dense, continuous network in which most atoms are bonded to three or more neighbours. Examples of covalent solids include diamond and graphite: brittle materials of great strength. Other examples include black phosphorus and germanium, which are of little interest as structural materials. In nanomechanical engineering, interest focuses on the more diamondlike covalent solids.

Diamondoid covalent solids

Used narrowly, *diamondoid* refers to material consisting almost entirely of carbon with sp^3 bonding. As used here, *diamondoid* describes strong, stiff covalent solids with a dense, three-dimensional network of bonds. The diamondoid solids of most interest have compositions biased toward multivalent first-row elements that form good covalent bonds (boron, carbon, nitrogen, oxygen), but may make substantial use of the similar second-row elements (silicon, phosphorus, and sulfur), limited use of monovalent elements (hydrogen, fluorine, chlorine), and sparing use of other elements. Among structures fitting this description (which embraces diamond itself) are many that are strong, stiff, and highly stable. Pure-carbon structures falling within this broad definition have been widely discussed.

In the present context, diamondoid structures have a further advantage. With some restrictions on the choice of atoms and bonding patterns, they are reasonably well described by molecular mechanics models, enabling the computationally economical investigation of mechanical systems specified in atomic detail.

Mixed structures

The categories though useful, are not sharply bounded. Many materials have intermediate structures. Ceramics with bonding intermediate between ionic and covalent have been mentioned. Covalent-solid objects of small size can be described as molecules, and combinations of them can be viewed as molecular or ionic solids. Covalent bonding can coexist with metallic conductivity. Molecular objects built on the development pathways from solution chemistry to advanced mechanosynthesis are likely to include extensively crosslinked polymeric structures intermediate between conventional polymers and covalent solids.

Materials vs. Molecular Structures

In macromechanical engineering, components are described as regions occupied by a material (usually isotropic and homogeneous) having certain mechanical properties: strength, density, shear modulus, and Young's modulus. These and other material properties (e.g., fatigue life, thermal coefficient of expansion, damping coefficients, friction coefficients), together with shape, enable a detailed engineering analysis of the component.

From a molecular mechanics perspective, a diamondoid component is a network of bonded atoms. The smallest objects contain few atoms, and must be described in atomic detail, yet at some scale a description in terms of shape and continuum material properties becomes quite useful (especially in preliminary design). To what extent can continuum models be applied to small diamondoid structures?

Bounded Continuum Approach

In extending continuum models to nanostructures, one must consider both surface effects and the atomic graininess of matter. Surface effects in diamondoid structures can be approximately described by a bounded version of a continuum model, but atomic structure constitutes a fundamental limit to such models. Surface effects on the mechanical properties of metals are more pervasive because of the nonlocal nature of metallic bonding; these effects are not considered here.

Surface effects

A simple continuum model treats the mechanical properties of a material as continuous up to a sharp boundary, but this is physically unrealistic for nanoscale objects. Models of interacting diamondoid surfaces that account for van der Waals attractions and overlap repulsions between their surface atoms, yet treat those atoms as forming a twodimensional continuum. This approach can be used to convert a standard continuum model into a ***bounded continuum model***, treating the boundaries of an object in a physically realistic way.

To apply a continuum model to the interior of an object, however, a sharp boundary must be defined. The natural definition of a surface for mechanical purposes—as the boundary of the region from which other similarly defined objects are excluded—gives misleading results in calculations based on bulk density, modulus, and strength. At ordinary pressures, nonbonded-contact radii are approximately twice as large as covalent radii, hence surface atoms occupy more space than interior atoms. Accordingly, rules for defining surface locations that yield good values for mass (from density), for component stiffness (from modulus), or for component strength (from material strength) define surfaces that fall well within the boundaries of the excluded volume.

Graininess effects

The atomic structure of matter constrains the shapes possible for small objects. In macroscale engineering, one assumes that (save for machining geometry and minimum gauge limitations) a part can be made in any desired shape, within some size tolerance and surface roughness.

Anisotropies

The bounded continuum models used in this volume treat materials as isotropic. Although this is a good approximation for many structures, including the most diamondlike diamondoids, it is a poor approximation for others. Continuum models of anisotropic bulk materials (e.g., those developed to describe composite structures) can be useful in describing anisotropic diamondoid components.

SURFACE EFFECTS

Density, modulus, size, and shape are ambiguous in nanoscale systems. Size and shape are ambiguous because molecular surfaces are not sharply bounded, and this makes density and modulus ambiguous, despite the clear-cut meanings of mass and stiffness. Nonetheless, a model based on defined surfaces and (isotropic, homogeneous) properties can provide useful design-phase estimates of mass, stiffness, and strength. A conservative application of bounded-continuum estimates can be used to identify design conditions that limit elastic deformation to acceptable bounds, but may provide only a rough description of the residual displacements.

Materials and Stiffness

To maximize strength and stiffness, nanoscale structural components are best made from diamondoid solids. In the systems analyzed in the following chapters, stiffness requirements are usually more constraining than are strength requirements. Stiffness can be directly estimated by computational experiments using molecular mechanics models.

Assigning sizes

Ordinarily, one regards the region occupied by a component as the region from which other components are excluded. This measure of size varies with compressive loads, and with the shapes and chemical natures of the facing surfaces. A reasonable standard choice is the region occupied by the component atoms of he structure when each is assigned a summable 0.1 nN radius; these radii are used in space-filling molecular representations throughout this volume.

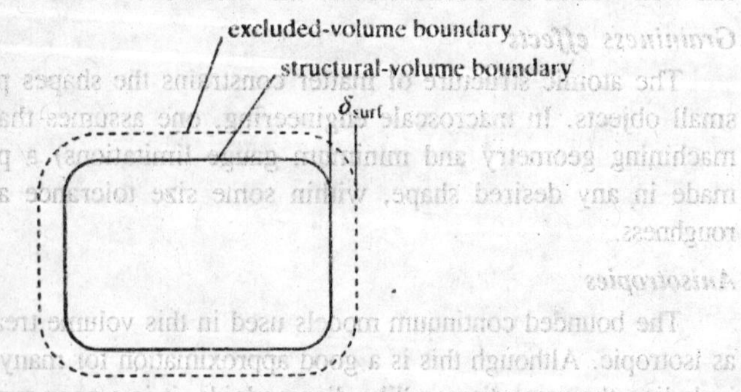

Fig. 10.1. Excluded vs. structural volume. The nonbonded-contact radii of surface atoms define the excluded volume of a diamondoid component.

The polycyclic *framework structure* accounts for most of the stiffness, but the excluded volume extends well beyond this framework, because of both the separation between nonbonded atoms and (in some structures) the presence of monovalent surface atoms. To maximize stiffness for a given component size, it is best to avoid monovalent surface atoms, instead favouring structures. The size correction δ_{surf} required to describe the structural volume is then approximately the difference between the covalent and nonbonded radii; using the 0.1 nN nonbonded radii, $\delta_{surf} \approx 0.07$ nm for both first- and second-row atoms. A larger correction ($\delta_{surf} = 0.1$ nm is used) yields a more conservative estimate of stiffness for a component of a given excluded-volume size.

A further correction can arise from differences in the stiffness contributions between the outermost framework atoms and those in the interior; with fewer constraints on their motions, they may be subject to greater relaxation under stress, reducing their stiffness contributions. This effect can be explored by examining the scaling of stiffness with cross-sectional area in a molecular mechanics model such as MM2.

Computational Experiments on Rod Modulus

Molecular mechanics predictions for the linear modulus of a rod, E_l (having the dimensions of force), can be obtained by computing the energy of a series of rods placed under differing strains. For an ideal measure of area S_{rod}, the product $S_{rod}E = E_l$, for some value of Young's modulus E.

A series of segments from rods with differing cross-sectional areas, most consisting of a hydrogen-terminated region of the diamond lattice. For these rods, a reasonable measure of cross-sectional area is n_{bonds}, the minimum number of framework bonds crossing a surface that divides the rod: (a), a strand of polyethylene, $n_{bonds} = 1$; for the two joined strands of (b), $n_{bonds} = 2$; and so forth, through $n_{bonds} = 9$. Calculating the effective area of these rods from their bond count and from the bond density for the equivalent orientation in diamond ($\sim 1.925 \times 10^{19}$ m^{-2}) avoids the issues of surface definition just discussed. For this family of rods

$$S_{rod} \approx 5.2 \times 10^{-20} n_{bonds} (m^2) \qquad ...(1)$$

and, in the bulk-material limit,

$$E_\ell \approx 5.45 \times 10^{-8} n_{bonds} (\text{N}) \qquad ...(2)$$

If relaxation were to reduce the stiffness contributions of surface atoms, then thinner rods would have a lower value of E_l. As can be seen, on a per-bond basis, thinner rods have a slightly *greater* value of

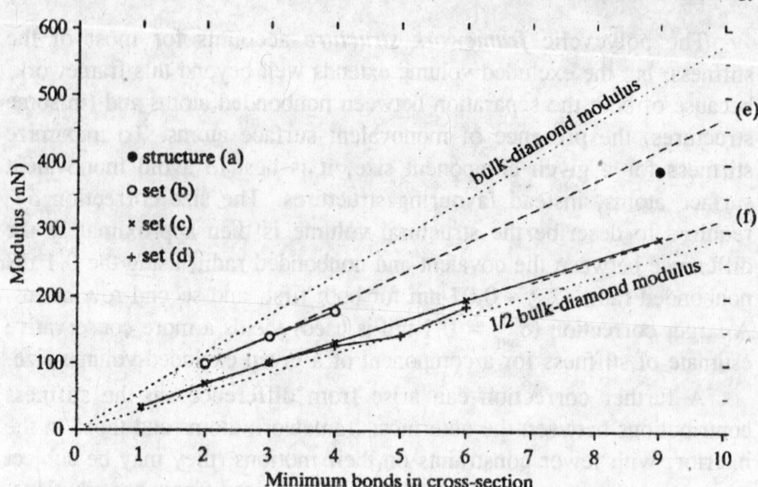

Fig. 10.2. Results of computational experiments measuring rod moduli in the MM2/CSC approximation.

E_l than thicker rods (presumably owing to contributions from nonbonded interactions involving hydrogen), but rods of all sizes have a value substantially lower than that of bulk diamond. Since this trend continues to structures with substantial stiffness contributions from interior atoms (e.g., the n_{bonds} = 18 value), it appears that this modulus deficit results from defects in the MM2 model rather than from surface relaxation.

With N and O termination, rather than CH and CH_z. (Chains of Sp^3 nitrogen atoms should be stable when each is bonded to carbon in a diamondoid solid of this kind.) For a given value of n_{bonds}, structures terminated in this manner are substantially more compact, and the MM2 model predicts substantially greater stiffness. This increase is a surface effect, but one which it is conservative to neglect. The hexagonal column (n_{bonds} = 3) is the smallest structural unit that might plausibly exhibit the elastic properties of bulk hexagonal diamond along the corresponding axis; it falls almost precisely on a line drawn through the result for a compact column of hydrogen-terminated hexagonal diamond structure with n_{bonds} = 16. Accordingly, surface relaxation effects on modulus appear negligible in this family of structures as well.

Shape Control in Irregular Structures

Nanomechanical devices can be specified in complete atomic detail, permitting them to be described (with varying accuracy) by molecular modeling techniques. This is necessary for small structures, and for structures about to be built, but it is not always necessary for large structures during preliminary design and analysis. It will be useful to have some guidelines regarding the boundary between small and large for these purposes.

Control of Shape and Detail of Specification

Nanomechanical engineering differs from macroscale engineering in that the number of possible component shapes can be significantly constrained by atomic sizes and feasible bonding patterns. In sufficiently small components, the set of possible shapes can be exhaustively described. For each type of mechanism, there is some minimum size below which components of the required shapes cannot be constructed (e.g., it is fruitless to attempt to build a gear having 50 teeth in a volume that can contain only 10 atoms; the limiting size for a 50-tooth gear is substantially larger). Near this minimum size, a design must be specified in complete molecular detail to gain any confidence in its validity.

Somewhat above this limit, however, complete atomic specification becomes unnecessary, just as in macroscale engineering and for the same reason: anyone of many arrangements of atoms can serve the purpose. On this larger size scale, specific molecular structure must be chosen before planning a mechanosynthesis and before performing a fully detailed analysis, but this choice can be made after performing a reasonably accurate analysis describing size, shape, strength, stiffness, mass, dynamics, energy dissipation, and performance. It has already been discussed how the number of available diamondoid structures varies with volume and structural constraints, drawing conclusions regarding the scale at which molecular detail can be omitted and a bounded continuum model can be adopted. Designs based on these models often are relatively conservative, setting only lower bounds on the performance of nanomechanical systems.

Estimates of the Number of Diamondoid Structures

Exponential scaling with volume

The number of distinct diamondoid structures N_{struct} that can fit in a given volume of reasonable shape is an exponential function of the volume (neglecting surface corrections). This can be seen by considering a hypothetical atom-by-atom construction process: at each surface site in a partially completed structure there is some number of distinct options (with a geometric-mean value N_{opt} which is not necessarily an integer) for how to proceed at the next step, that is, for what kind of atom to bond in what manner.

The value of N_{opt} describes the effective number of choices made per atom added, correcting for patterns of choices that in the end produce the same structure. The number of choices defined in this way is proportional to the number of atoms in the final structure (which will, for some typical atomic number density n_v be proportional to the volume V), and each choice leads to a different structure (by the above definition of *distinct choice*). For a suitable definition of N_{opt}, the number in question is

$$N_{struct} = N_{opt}^{(n_v V)} \quad ...(3)$$

The proper value of N_{opt} is not obvious, even if the atoms are restricted to a single type. It will depend on the range of permissible local structures, constrained by (for example) the magnitudes and patterns of bond strain permitted in the interior of a solid stable enough for practical use. Knowledge of these constraints would be necessary (yet not sufficient) to enable a simple computation.

Estimates from entropy of fusion

The entropy of fusion of a crystal is a measure of the increase in the available volume in configuration space that occurs with a transition from a single, regular structure to an ensemble of states which includes many different structures. If the potential wells in the liquid state were as narrow as those in the solid state, and if each of those potential wells were equally populated and corresponded to a stable amorphous structure (and vice-versa), then the entropy of fusion would be a direct measure of the increase in number of wells, and would thus provide a direct measure of the number of available structures. In practice, low-stiffness wells and the thermal population of unstable transition structures tend to make the entropy of fusion an overestimate; the negligible thermal population of high-strain states that would be kinetically stable in corresponding solids tends to make it an underestimate. The absence of a choice of atom types at various sites (in elemental crystals) introduces a strong bias toward underestimation, relative to the systems of interest here.

Measuring the entropy of fusion of carbon (melting point ~3820 K) presents technical difficulties; the values for silicon and germanium crystals, which share the diamond structure, are 4.6×10^{-23} and 4.7×10^{-23} J/K per atom respectively. These values are both about $3.4k$; if this increase in entropy were solely the result of the increased number of available potential wells per atom, that number would be $N_{opt} \approx 30$. An estimate of the number of distinct states in argon at liquid density yields a value equivalent to $N_{opt} \approx 3$ wells per atom; this is compatible with an estimate based on the entropy of fusion.

Estimates including multiple elements

In the set of broadly diamondoid structures built up from C, Si, B, N, P, 0, S, H, F, and Cl, N_{opt} is increased as a result of the choice of atom type at each step; a factor of 5 is a plausible estimate of this increase, allowing for constraints and biases in the choices made. Taking the preceding silicon and germanium estimates as a base, this suggests that a reasonable estimate for diamondoid structures is $N_{opt} \approx 150$, and that calculations based on $N_{opt} = 30$ are conservative. This conservatism is intended to compensate for various constraints, including structural stability and avoidance of unacceptable electric fields and potentials resulting from patterns of aligned dipoles.

A region with a structural volume of 1 nm^3, $N_{opt} = 30$, and $n_v = 100$ nm^{-3} has $N_{struct} \approx 10^{148}$. A similar cubical region with an excluded volume of 1 nm^3 and $\delta_{surf} = 0.1$ nm has a structural volume of ~ 0.51 nm^3, and $N_{struct} \approx 10^{75}$.

Geometrical Constraints

These enormous numbers imply that objects can be constructed with almost any specified shape, *provided* (1) that the specifications do not describe complex contours on a subnanometer scale, or sharp corners and the like, and (2) that the specifications allow adequate tolerances in the placement of surface atoms, relative to the ideal contours of the shape. Demanding that many surface atoms simultaneously meet tight tolerances can easily result in an over-constrained problem having no physically realizable solution.

For example, one might ask that each of the ~ 100 atoms on the surface of a 1 nm object be within a distance e of an ideal surface contour. A randomly chosen structure can be terminated so as to keep only those atoms within a certain boundary, and that boundary can be chosen so that the surface atoms of the terminated structure are close to a preselected ideal surface contour. With $n_v = 100$ nm^{-3}, the positions of these surface atoms typically vary over a range of ~ 0.2 nm about that contour. If we demand that each surface atom be within $\varepsilon = 0.01$ nm of this contour, then on the order of $(0.01/0.2)^{100} \approx 10^{-130}$ of a set of randomly chosen structures will meet this condition. Given $N_{struct} = 10^{75}$, the probability that some structure satisfies this condition is on the order of 10^{-55}. If, however, one asks that 50 atoms forming particular surfaces be accurately placed within $\varepsilon = 0.02$ nm, then on the order of $(0.02/0.2)^{50} \approx 10^{-50}$ of a set of randomly chosen structures will meet this condition, and the probability that a suitable structure does *not* exist is on the order of 10^{-25}.

Tight constraints on object shape arise chiefly in the working interfaces of moving parts, for example, where surfaces must slide smoothly. Several models for such interfaces; some are based on the choice of regular structures of high symmetry (to which the statistical analysis in this section does not apply); others describe the constraints on an irregular structure imposed by the requirement that it slide smoothly over a regular surface. Other working interfaces bind molecules. Nonworking surfaces are exposed to free space and have few constraints.

Molecular Binding

Surfaces with binding sites (receptors) for specific molecules are useful, for example in acquiring and processing molecules from solution. The principles of selective binding are familiar from molecular biology, and chiefly involve detailed surface complementarity between the molecule and its receptor: complementary surface shapes to provide strong van der Waals attraction, complementary patterns of charge and dipole orientation to provide electrostatic attraction, and so forth (in aqueous solutions, so-called hydrophobic forces can playa large role). These requirements for complementarity impose constraints on surface structure, again reducing the size of the set of acceptable structures by many orders of magnitude.

Mammalian immune systems provide an example in which structures capable of strong, selective binding are chosen from a set of structures of calculable size. Mammals can produce highly specific antibodies able to bind anyone of a wide range of small molecules (when a molecule of a particular type is presented as an immunologically stimulating *hapten*); this includes molecules not found in nature. These antibodies are developed by a process of variation and selection chiefly within a space of possibilities defined by differing sequences of the 20 genetically encoded amino acid residues within the antibody hypervariable domains (variations at other sites are relatively rare). These domains contain ~38 amino acid residues; hence most antibodies are selected from a set of structures containing $\sim 20^{38} \approx 3 \times 10^{49}$ members. Equation (3) with $N_{apt} = 30$ suggests that comparable diversity can be provided by a set of structures each containing <35 atoms, although the geometric and structural constraints of receptors often require larger structures.

Kaehler Brackets

T. Kaehler has proposed a systematic approach to designing irregular interfaces to meet particular geometrical constraints. First, one defines an anchoring surface [e.g., a hydrogenated diamond (111) surface]. Second, one defines a large family of brackets extending from this surface. But adding various modifications: substitution of N, B, P, or SiH for CH, replacement of a C—C linkage or a CH|HC contact with C—CH_2—C, C—O—C, or C—S—C, etc., all subject to stiffness and stability constraints. The size of the resulting family of structures increases exponentially with bracket length and soon reaches millions. Third, one determines the geometry of each bracket and notes the positions and orientations of its terminal bonds. Finally, one compiles an index based on this data, organized to enable one to find the brackets with terminal bonds most nearly matching a desired position and orientation.

How might an indexed set of Kaehler brackets be used in design? The brackets define a relatively comprehensive set of slim diamondoid structures, indexed according to the geometrical relationship between their ends. These can be used link structures that must be held in specific relative orientations; several such bridges can form a stiff connection.

Further, the design-phase substitution of one Kaehler bracket for another bracket of similar end geometry is like moving small robotic arm by a small distance (and perhaps changing its stiffness). One can imagine algorithms that in effect push and pull a flexible structure into a desired shape (e.g., to fine-tune a receptor design), choosing brackets of the shape necessary to apply the needed forces. Discrepancies between predicted and actual geometries could be rectified by further substitutions among brackets of similar end geometry, but perhaps quite dissimilar structure.

High Rotational Symmetry

In building rotating machinery, components having high-order rotational symmetry are often desirable. Rotationally symmetric components divide roughly into two classes: *strained-shell* structures made (at least conceptually) by bending a straight slab into a hoop, and *special-case* structures that are more intrinsically cylindrical. Members of an intermediate class of *curved-shell* structures resemble strained-shell structures, but with regular arrays of dislocation-like features that reduce the strains. These terms correspond to rough categories, not to precise distinctions.

Strained-shell Structures

A relatively small strained shell having two distinct layers of cyclic structure. The thickness-to-radius ratio (*t*/*r*) of strained shells is constrained by the net hoop stress and the permissible bond tensile strain at the outer surface. For diamondoid structures with negligible net hoop stress, permissible bond strains may be as large as 0.035 nm ($\sim 1.23\ r_0 \approx 0.187$ nm); assuming a Morse potential, a C—C bond with smaller strain has positive stiffness, tending to stabilize its length.

In the presence of a strain gradient perpendicular to the axis of strain (as in a strained shell), even a layer somewhat beyond the positive-stiffness limit can be stabilized by restoring forces imposed by the less-strained layers closer to the shell axis (the layers are coupled by shear stiffness). *Note that this situation differs from the single-bond, pure-tensile-load model used to derive bond failure rates.* In general, the stability analysis for highly strained shells is complex, because of the possibility of concerted bond rupture and rearrangement; the designs in this volume assume considerably smaller bond strains.

If the material of the shell had linear elastic properties, a 0.035 nm strain would occur with $t/r \approx 0.46$ (using a mid-thickness reference for *r*). The worst-case nonlinearity would prohibit bond compression entirely, resulting in a limit of $t/r \leq \sim 0.23$. Both these estimates neglect the favourable effects of angle-bending relaxation. The appropriate measure of thickness for these calculations is smaller than that of the structural volume; atomic centers at the edge of the framework structure provide a conservative marker for the surface.

R. Merkle (1991; 1992a) has developed computer-aided design (CAD) software able to generate a wide range of strained-shell structures based on the diamond lattice, providing flexible control of radius, thickness, crystallographic orientation, surface shape, and surface termination; precise arrays of thousands of atoms can be generated by supplying a few parameters. The structure was generated using this CAD tool together with subsequent energy minimization using molecular mechanics software.

Curved-shell Structures

Thick cylindrical structures have many applications. They can be made from diamondlike structures, relieving strain with arrays of dislocations. Merkle notes that the Lomer dislocation is stable and has useful geometrical properties for purpose. Preliminary work by R. Merkle and L. Balasubramaniam has produced software that aids in the design of tubular structures of this kind. Merkle suggests that computer-aided design tools able to automatically generate diamond lattices modified by (potentially dense) arrays of dislocations offer a promising approach to the generation of diamondoid structures of considerable diversity.

Special-case Structures

Sorme cylindrical structures are not members of scalable families. The rods are examples of small-radius structures having significant rotational symmetry. Similarly the cylindrical structure provides an example of intermediate size; the outer ring of the bearing provides a still larger example. The set of diamondoid structures of high rotational symmetry, small length, and diameter ≤1 nm is large, but may be small enough to permit the development of an exhaustive catalogue.

ADHESIVE INTERFACES

Large structures can be built directly from small reactive moieties and clusters, but there are substantial advantages to building indirectly via intermediate-scale blocks. These blocks can be joined by *adhesive interfaces* that can exploit any of the attractive forces responsible for the cohesion of solids. Designing a system to be assembled in this way imposes some constraints on its structure, but in many instances these need not be severe.

Van der Waals Attraction and Interlocking Structures

The simplest adhesive interface consists of two surfaces of complementary shape, with no special provision for bonding. If the interface has suitable corrugations, the interface will resist shear. Van der Waals attraction provides tensile strength and (together with balancing overlap repulsions) interfacial stiffness. A simple contact between complementary diamondoid surfaces can have a tensile strength ~1 GPa and a stiffness per unit area of >30 N/m·nm². The compliance of an interface with this stiffness is equal to that of a ~30 nm thick slab of diamond.

Introducing a simple van der Waals interface thus degrades the strength of a diamondoid structure by about two orders of magnitude, and substantially degrades the stiffness of systems built on a 100 nm scale. Where strength is unimportant and stiffness is not at a premium, such interfaces may be of considerable use. Convoluted, interlocking interfaces (e.g., with knobs that snap into holes) can greatly increase strength and stiffness by making greater use of overlap repulsions.

Ionic and Hydrogen Bonding

Complementary patterns of charge or hydrogen bonding capacity can increase interfacial strength and stiffness. The strength and stiffness of interfaces with extensive ionic bonding can be estimated from the properties of sodium chloride. Theoretical calculations of the breaking stress yield estimates around 2.7 GPa, and modulus data indicates an interlayer stiffness per unit area of ~ 170 N/m·nm^2.

Covalent Interfacial Bonding

The strongest and stiffest adhesive interfaces result from the formation of dense arrays of strong bonds; various reactive diamondoid surfaces can be designed to join in this way. Diamond (111) surfaces terminated with complementary patterns of Band N atoms can join by forming an array of dipolar bonds. Arrays of dienes and dienophiles can join through Diels-Alder reactions to form an array of C—C bonds. Hexagonal diamond surfaces can apparently bond to form continuous hexagonal diamond (bonding two planar surfaces is electronically analogous to bonding a hexagonal column to a surface); a pair of diamond (110) surfaces can apparently bond to form continuous cubic diamond. In these instances, adhesive surfaces can unite blocks of the strongest, stiffest materials known to experimental science without leaving a distinct interface, and hence without loss of strength or stiffness.

A potential difficulty in forming strong interfaces is avoiding the sudden release of unacceptable amounts of mechanical energy: the energy of a diamond (111) surface is ~ 5.4 J/m^2, which if suddenly thermalized would be enough to raise the temperature of a 1 nm thick slab by $\sim 10^3$ K. Adhesion, however, can be viewed as the inverse of crack formation. The gradual application of inhomogeneous stress distributions can cause controlled crack growth; the relaxation of similar stress distributions can enable controlled joining. Alternatively, interposed structures could force joining to proceed at a controlled rate, or surface structures might be chosen such that adhesion is slowed by steps with substantial thermal activation barriers.

CONCLUSIONS

Nanoscale structural components can serve either as housings or as moving parts. The best materials for nanoscale components are diamondoid in structure: these materials have unusual strength and stiffness, and lend themselves to design using molecular mechanics methods. Computational experiments indicate that bulk-material stiffnesses for diamond (and diamondoid solids) can be used to describe components of subnanometer dimensions, provided that the modulus is applied to a cross-sectional area that is descriptive of the structural framework (here termed the *structural volume*), omitting an outer layer of the *excluded volume* resulting from the relatively long range of overlap forces, and from any monovalent surface atoms. At a ~1 nm size scale, the number of stable diamondoid structures becomes enormous, and essentially any shape can be built, provided that tight (<0.02 nm) tolerances are not applied too widely. Components of high rotational symmetry can be constructed in several ways.

Some open problems

Open problems in the field of nanoscale structural components divide into problems of design and problems of analysis. Regarding the latter, further research using molecular mechanics studies to improve bounded continuum models and explore their range of validity would be valuable. A challenging objective is to develop software that merges classical finite-element methods with bounded continuum models and molecular mechanics methods; this would enable a designer to control the level of detail in a structural description, keeping atomic detail where it is significant and omitting it elsewhere to improve computational efficiency.

In design, it would be useful to specify and characterize many small, regular structures useful as shafts, gears, and so forth; means for indexing and recovering designs are of comparable importance. Still more useful would be the development of software for designing irregular structures that meet particular boundary conditions. The Kaehler bracket approach is promising for a range of applications, as is Merkle's suggestion of dense dislocation networks; another might be a simulated annealing process (using a potential function developed for the purpose) in which atom number and atom types are permitted to change.

In another area entirely, studies of the joining of diamond surfaces could resolve questions regarding the practicality of using the strongest adhesive interfaces in component assembly. Further, characterization of the properties of B-N dipolar bonds (and other dipolar and coordination bonds) could yield descriptions of desirable adhesive interfaces, and might yield improved designs for interfaces suitable for use in nanomechanical gears.

INDEX